REVISE AQA GCSE
Science
Further Additional Science
REVISION WORKBOOK

Series Consultant: Harry Smith   Authors: Iain Brand and Peter Ellis

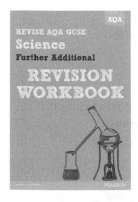

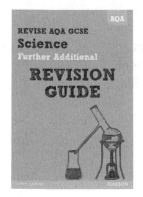

THE REVISE AQA SERIES
Available in print or online

Online editions for all titles in the Revise AQA series are available Spring 2013.

Presented on our ActiveLearn platform, you can view the full book and customise it by adding notes, comments and weblinks.

Print editions

Further Additional Science Revision Workbook   9781447942160

Further Additional Science Revision Guide   9781447942498

Online editions

Further Additional Science Revision Workbook   9781447942269

Further Additional Science Revision Guide   9781447942276

Print and online editions are also available for Science (Higher and Foundation) and Additional Science (Higher and Foundation).

This Revision Workbook is designed to complement your classroom and home learning, and to help prepare you for the exam. It does not include all the content and skills needed for the complete course. It is designed to work in combination with Pearson's main AQA GCSE Science 2011 Series.

> **To find out more visit:**
> www.pearsonschools.co.uk/aqagcsesciencerevision

ALWAYS LEARNING   PEARSON

Contents

1-to-1 page match with the Further Additional Revision Guide ISBN 9781447942498

 Anything with a sticker like this is helping you to apply your knowledge and practice your skills.

A small bit of small print

Target grade ranges are quoted in this book for some of the questions. Students targeting this grade range should be aiming to get most of the marks available. Students targeting a higher grade range should be aiming to get all of the marks available.

Target grade ranges

AQA publishes Sample Assessment Material and the Specification on its website. This is the official content and this book should be used in conjunction with it. The questions in this book have been written to help you practise what you have learned in your revision. Remember: the real exam questions may not look like this.

Into and out of cells

G-E **1** Draw a ring around the correct answer to complete the following sentences.

(a) Substances that move into and out of cells must be

> dissolved
>
> ions
>
> solid

.

(1 mark)

(b) In diffusion, the net movement of molecules is

> down a concentration gradient
>
> up a concentration gradient
>
> both ways across a concentration gradient

.

(1 mark)

D-C **2** Osmosis can be shown in a laboratory in an experiment using Visking tubing.

(a) Give a definition of **osmosis**.

> **Guided**

Osmosis is the movement of .... molecules across a ....

.... from areas of .... to areas

of .... *(3 marks)*

> On every page you will find a guided question. Guided questions have part of the answer filled in for you to show you how best to answer them.

(b) The diagram shows part of this experiment.

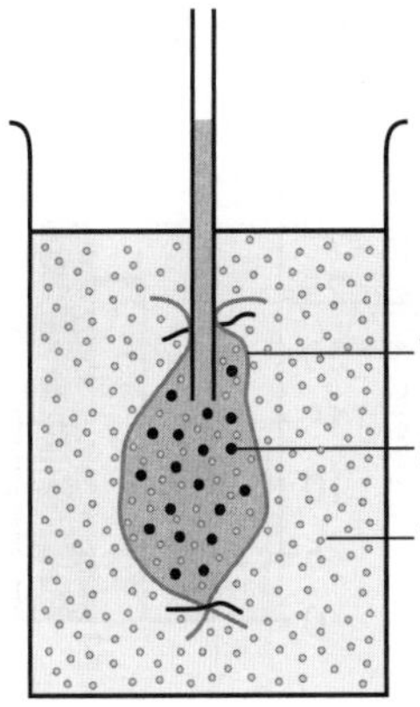

(i) Describe how the rate of osmosis into the Visking tubing could be measured.

....

.... *(1 mark)*

(ii) Explain why the sugar molecules do not move in this experiment.

....

....

.... *(2 marks)*

Visking tubing

sugar

water

B-A* **3** Plants need to take in minerals, which are in very low concentrations in the soil.

(a) Name the process that plants use to take in minerals.

.... *(1 mark)*

(b) Explain why plants use this process.

> Your explanation should cover why plants use this process rather than diffusion.

....

....

.... *(2 marks)*

(c) Explain why the plant needs to take in oxygen to make this process happen.

....

.... *(2 marks)*

Sports drinks

G-E

Guided

1 A gym sells a sports drink to its customers. The drink contains a mixture of sugars.

What **two** other ingredients is the drink likely to contain?

The drink is mostly composed of ....

but will also contain dissolved .... *(2 marks)*

D-C

2 Explain why the proportions of ingredients are important in a sports drink.

....

....

.... *(3 marks)*

B-A*

3 A manufacturer has developed a new drink called SportAde.

(a) The manufacturer describes the drink as **isotonic**. This means that the concentration of sugars and ions in the drink are the same as in the body's tissues.

Explain why an isotonic drink is likely to be better when exercising than drinking just water.

....

.... *(2 marks)*

AQA SKILL
Evaluate
Page 98

EXAM ALERT

(b) The manufacturer of the drink does some research into how long people can exercise when drinking SportAde compared with not drinking anything.

Legend: ☐ Time exercising without SportAde ■ Time exercising with SportAde

If you are given data in a question make sure you use it to answer the question.

Students have struggled with questions like this in recent exams – **be prepared!**

(i) Describe what the data show.

....

.... *(2 marks)*

(ii) The manufacturer of SportAde claims that the drink lets people exercise for longer. Use data from the graph to evaluate this claim.

....

....

.... *(3 marks)*

(iii) The manufacturer of the drink also claims that it helps people recover from exercise. How far do the data support this claim?

....

.... *(2 marks)*

2

Exchanging materials

G-E   **1**   Draw a ring around the correct answer to complete the following sentences.

(a) Alveoli in the lungs are useful in gas exchange because they

allow carbon dioxide to pass into the blood
have a thick membrane
provide a large surface area

.

(1 mark)

(b) The blood vessels that surround the alveoli are called

arteries
capillaries
veins

.

(1 mark)

D-C   **2**   Absorption of digested food molecules takes place in the small intestine. The small intestine has a surface adapted to assist this process.

⟩ **Guided** ⟩   (a) Describe how the small intestine is adapted to help to absorb food molecules.

The surface of the small intestine is covered with ....

These help by increasing .... *(2 marks)*

(b) Explain why these structures need to have thin walls.

....

....

.... *(2 marks)*

B-A*   **3**   All living things need oxygen to respire. Organisms like mammals use lungs to take in oxygen but bacteria use a different mechanism.

(a) Alveoli in the lungs have an efficient blood supply. Explain why it is important that the blood supply to the alveoli is both plentiful and kept flowing.

....

....

.... *(3 marks)*

(b) Organisms such as bacteria do not have lungs or a blood supply. Explain why bacteria do not need these features in order to take in oxygen from their surroundings.

Your answer should also mention the process that bacteria use to take in oxygen.

....

....

.... *(3 marks)*

Ventilation

G-E **1** Complete the following passage by using words from the box.

abdomen	carbon dioxide	glucose	heart	lungs	oxygen	thorax

The largest organs in our breathing system are the ....

These organs are in the .... of the body.

They let air into the body so that .... can be taken into the blood and

.... can leave.

(4 marks)

D-C **2** The diagram shows the human breathing system.

(a) Give the name of the part of the system labelled X.

.... *(1 mark)*

(b) The part of the system labelled Y helps with breathing, but has another role.

(i) Name this part.

.... *(1 mark)*

▷ **Guided** ▷

(ii) Explain the role that it has.

It helps to ....

organs in the thorax of the body, especially the

.... *(2 marks)*

B-A* **3** The body relies on ventilation for survival.

(a) Describe the gas exchange processes that happen during ventilation.

....

.... *(2 marks)*

(b) Explain how the ribs and diaphragm work together to allow ventilation to take place.

In the exam, you should know both processes – breathing in and breathing out. However, for the purposes of your revision answer here, concentrate on one of the two processes. Answers for both will be given – so you can do both and check that you can do both ways correctly!

....

....

....

.... *(4 marks)*

(c) Explain why artificial ventilators are useful.

....

.... *(2 marks)*

Exchange in plants

G-E **1** To make proteins and sugars, plants need to take in a variety of chemical substances. These substances include carbon dioxide and mineral ions.

(a) Name **one** other substance that plants need to take in. .... *(1 mark)*

Guided (b) Describe how plants take in carbon dioxide for this process.

Carbon dioxide enters the .... of the plant by the process of

.... *(2 marks)*

D-C **2** Describe how a plant is adapted to take in mineral ions efficiently.

....

....

.... *(3 marks)*

D-C **3** The diagram shows a cross-section of part of the leaf of a plant.

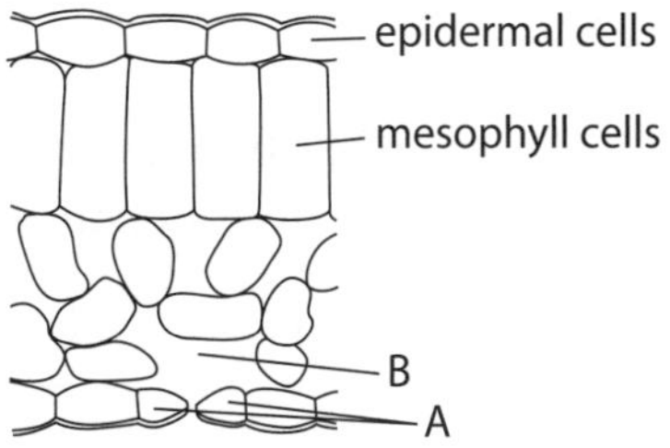

(a) (i) Name the part of the leaf labelled A.

....   *(1 mark)*

(ii) Describe the role of the part of the
leaf labelled A.

....

.... *(2 marks)*

epidermal cells

mesophyll cells

B

A

(b) Explain the function of the part of the leaf labelled B.

....

.... *(2 marks)*

B-A* **4** Some students investigated the rate at
which water evaporated from leaves using
this apparatus.

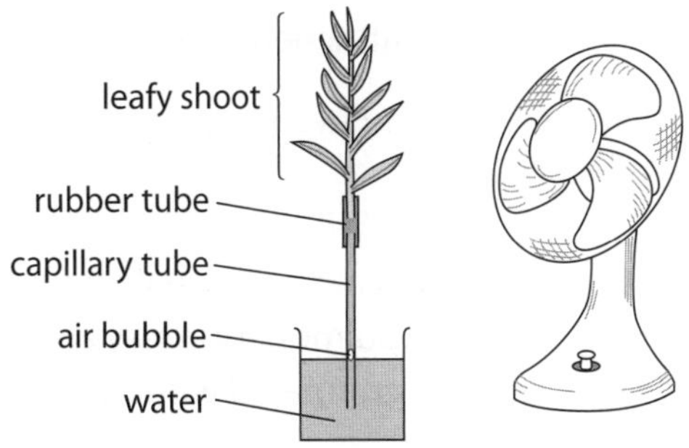

The students measured how far the air bubble travelled
up the capillary tube in five minutes with the fan on,
and with the fan off. They found that the bubble moved
9 mm with the fan off and 13 mm with the fan on.

leafy shoot

rubber tube

capillary tube

air bubble

water

(a) Explain the results the students collected.

....

.... *(2 marks)*

In this question, you need to think about what the plant does as the speed of the wind increases.

(b) The speed of the fan is increased, and it is found that the rate that the bubble moves does not increase. Suggest, in terms of the plant's response, why this is the case.

....

....

.... *(3 marks)*

The circulatory system

G-E 1 Draw a ring around the correct answer to complete the following sentences.

(a) The heart is part of an organ system in the body called the

circulatory system
digestive system
respiratory system

.

(1 mark)

(b) The heart is mostly composed of the same tissue – this tissue is

blood
epithelium
muscle

.

(1 mark)

D-C 2 The human heart is divided into two sides. Each side of the heart contains two chambers.

> **Guided**

(a) Name the **two** chambers found on the right side of the heart.

The two chambers are the right .... and the right .... *(2 marks)*

(b) These two chambers are separated by an important structure. Name this structure and state its function in the heart.

....

.... *(2 marks)*

B-A* 3 The diagram shows a section through the human heart.

Blood leaves the right side of the heart in the pulmonary artery and flows towards the lungs. The blood then returns to the left side of the heart in the pulmonary vein.

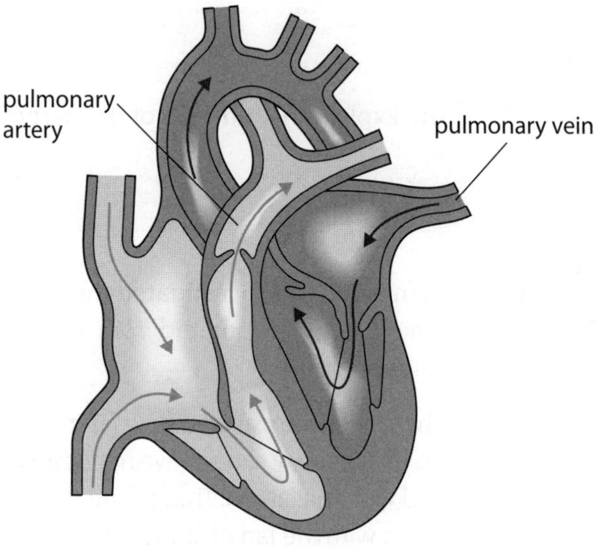

(a) Name the other **two** blood vessels attached to the heart.

....

.... *(2 marks)*

(b) What is unusual about the blood in the pulmonary artery and the pulmonary vein compared with other veins and arteries?

....

....

.... *(2 marks)*

(c) The heart needs to move blood. Suggest why this leads to the heart having a different thickness on each side.

> You need to consider where the blood on each side of the heart is going, and how the heart makes the blood move.

....

....

.... *(3 marks)*

Blood vessels

G-E   **1**   Blood is carried around the body by different types of blood vessel. Two of these types of vessels are arteries, like the aorta, and veins, like the vena cava.

     (a) Which **one** of these statements is correct? Tick (✓) **one** box.

        ☐   The vena cava carries oxygenated blood into the heart.

        ☐   The vena cava carries deoxygenated blood away from the heart.

        ☐   The aorta carries oxygenated blood away from the heart.

        ☐   The aorta carries deoxygenated blood into the heart.   *(1 mark)*

Guided   **(b)** Describe the structure of an artery.

        An artery has .... walls. These walls are composed of two types of fibres:

        .... tissue and .... fibres.   *(3 marks)*

D-C   **2**   Angina is a condition that causes chest pain. It happens when the coronary arteries supplying blood to the heart muscle become narrowed. Explain how angina could be treated.

     ....

     .... *(2 marks)*

D-C   **3**   Blood flowing into the organs of the body needs to penetrate into the organs to carry the blood to tissues.

     (a) Name the blood vessels that take blood to organs.

     .... *(1 mark)*

     (b) These blood vessels are different in structure to blood capillaries. Explain why these differences are important.

        ....

        ....

        ....

        .... *(4 marks)*

B-A*   **4**   Veins carry blood away from the organs of our body. Blood in the veins often has to travel long distances to return to the heart.

     (a) Explain how the structure of veins helps them carry blood for long distances.

        ....

        .... *(2 marks)*

     (b) A doctor taking blood from a patient will insert a needle into a vein to take blood. Suggest why a doctor takes blood from a vein and not an artery.

> There are two possible answers here – you need to consider either the structure of the different blood vessels, or else the way in which each transports the blood they contain.

        ....

        .... *(2 marks)*

Blood

G-E 1 Use words from the box to complete the sentences below.

heart	kidneys	liver	lungs	platelets	red blood cells	water

Blood plasma is mostly composed of .... The role of the plasma is to

transport materials around the body. For example, the plasma carries carbon dioxide to

the .... Another waste material is urea, which is carried in the blood

from the .... to the .... *(4 marks)*

D-C 2 The blood contains large numbers of red blood cells.

(a) Name the cell structure, normally present in animal cells, that is absent from the red blood cell.

.... *(1 mark)*

> **Guided** **(b)** Red blood cells contain a pigment that gives them their colour.

(i) Name this pigment and describe its role in the blood.

This pigment is called .... Its role is to combine with

.... in the lungs, and carry it around the body.

(2 marks)

(ii) Explain what happens in red blood cells when the blood reaches respiring tissues.

> **Explain** means that you have to describe what happens and then say **why** it happens.

....

.... *(2 marks)*

B-A* 3 Describe the role of **platelets** in the blood.

EXAM ALERT

....

....

....

....

....

> You should know the functions of all the different parts of the blood, including plasma and platelets.

> Students have struggled with questions like this in recent exams – **be prepared!**

.... *(3 marks)*

B-A* 4 White blood cells normally make up about 1% of the blood. A man goes to the doctor, complaining of flu-like symptoms. The doctor takes a blood test to look at the number of white blood cells in the man's blood. The number of white cells is greater than when the man was healthy. Explain why.

....

....

....

.... *(3 marks)*

Transport in plants

G-E **1** The diagram shows the apparatus that was used in an experiment to investigate the movement of water in the shoot of a plant. The experiment was carried out at 20 °C.

Circle A shows the position of an air bubble at the start of the experiment and circle B shows its position at the end.

Draw a ring around the correct answer to complete the following sentences.

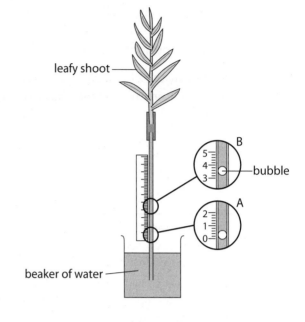

leafy shoot

B

bubble

A

beaker of water

(a) The distance travelled by the water bubble

during this experiment is
| 3.2 cm |
| 3.5 cm |
| 3.7 cm |
.

(1 mark)

(b) The movement of water along shoots in living plants is called the
| osmotic stream |
| photosynthesis stream |
| transpiration stream |
.

(1 mark)

D-C **2** Plants have two types of specialised tissues used for transport.

(a) One of these tissues is the xylem. Describe the role of **xylem tissues**.

The xylem is used to transport .... and ....

in the plant from .... to ....   *(3 marks)*

(b) Dissolved sugars are transported in the other type of tissue.

(i) What is the name of this type of tissue?

....   *(1 mark)*

(ii) Describe what happens to the sugars that are transported by this tissue.

> The sugars can be transported for two different purposes. Make sure your answer mentions both.

....

....   *(2 marks)*

B-A* **3** Plants rely on the transpiration stream to stay alive. Explain how the transpiration stream arises in plants and why it is so important.

....

....

....

....   *(4 marks)*

9

Biology six mark question 1

Human blood contains different types of cells, suspended in the plasma. The pie chart shows the typical composition of a sample of blood.

Explain why the blood needs all these different components in order to keep humans alive and healthy.

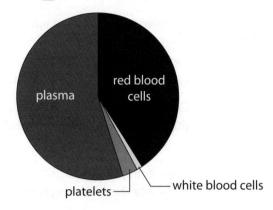

You will be more successful in six mark questions if you plan your answer before you start writing. In this question, some of the work has been done for you – the names of the different components of the blood have been provided.

Your answer should include:
- a basic description of the function of each component of the blood
- how each part of the blood helps the body to work normally.

Remember that the examiner will also be looking at your quality of written communication for this question. Therefore, you should think about the best way to present your answer. A long paragraph of text might be difficult to read, and you may miss out one component in your answer. You may want to write your answer using sub-headings for each component of the blood.

....

....

....

....

....

....

....

....

....

....

....

....

....

....

....

....

.... *(6 marks)*

Removing waste products

G-E **1** Draw a ring around the correct answer to complete the following sentences.

(a) One of the waste products of respiration is
| carbon dioxide |
| glucose |
| oxygen |
.   *(1 mark)*

(b) This waste product is removed from the body through the
| kidney |
| lungs |
| skin |
.   *(1 mark)*

D-C **2** The kidney is an organ that removes waste products from the body.

(a) How do waste products reach the kidney?

.... *(1 mark)*

⟩ **Guided** ⟩ **(b)** Describe the difference between urea and urine.

Urea is produced in the .... by the breaking down of

....

Urine is produced in the .... and is stored in the

.... before being excreted from the body.   *(4 marks)*

B-A* **3** The table shows the composition of the fluid being filtered in the kidneys, both before and after passing through a person's kidneys.

Substance	Concentration in blood vessel entering the kidney	Concentration in blood vessel leaving the kidney
glucose	high	high
urea	high	low
ions	high	low

(a) Describe how urine is produced in the kidneys.

....

.... *(2 marks)*

(b) Use the data in the table to decide whether this person's kidneys are healthy and working properly.

| You should consider the changes in the substances as the blood moves through the kidneys. |

....

....

.... *(3 marks)*

(c) An unhealthy kidney does not absorb enough water. Explain why this may cause health problems.

....

....

.... *(3 marks)*

Kidney treatments

G-E   **1**   The chart shows the employment status of people who have had different forms of treatment for kidney failure.

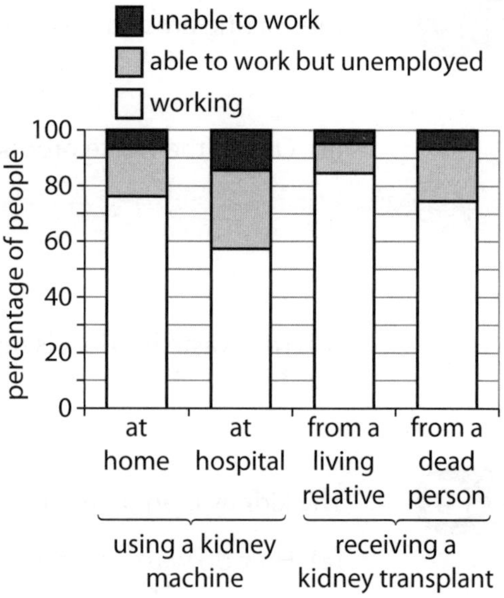

   (a) Use the graph to estimate the percentage of people who use a kidney machine in hospital and are unable to work.

   ....

   .... *(1 mark)*

   (b) Use the graph to state which form of treatment allows the largest percentage of people to be able to work.

   ....

   .... *(1 mark)*

Legend:
■ unable to work
▨ able to work but unemployed
☐ working

percentage of people (y-axis: 0, 20, 40, 60, 80, 100)

x-axis categories:
at home | at hospital (using a kidney machine) | from a living relative | from a dead person (receiving a kidney transplant)

D-C   **2**   Look at the data in question **1** for people receiving treatment with a kidney machine. Suggest why the place where a kidney machine is used can affect a patient's employment status.

....

....

.... *(3 marks)*

D-C   **3**   People with kidney failure can be connected to a dialysis machine, which carries out a similar function to the kidneys.

   (a) Describe how the dialysis machine works.

....

....

.... *(3 marks)*

Guided   **(b)** In a dialysis machine, the dialysis fluid used contains glucose and mineral ions isotonic with the blood. Why is it important that dialysis fluid contains these substances, rather than being water?

   If water were used as dialysis fluid, then the patient would lose ....

   from their blood via the process of ...., and take in water from the

   dialysis solution by .... *(3 marks)*

B-A*   **4**   One possible treatment for people who have a damaged or faulty kidney is to have a kidney transplant from a healthy person. It is possible for kidneys to be transplanted from 'living donors' because, although humans have two kidneys, it is possible to live quite normally with only one.

AQA SKILL
Evaluate
Page 98

   Evaluate the use of transplants as a form of treatment for kidney disease.

....

....

....

.... *(4 marks)*

Body temperature

G-E **1** Use words from the box to complete the sentences below.

blood	brain	food	skin	sweat	water

The temperature of the body is monitored and controlled by the .... If the body

becomes too hot, then it starts to produce .... When this happens, the body loses

........................., which will need to be replaced either in our drink or in our ....

(4 marks)

D-C **2** The average temperature during the day in January in the UK is around 3 °C. If you go outside without warm clothing your body temperature will start to drop.

(a) How will the body detect a decrease in skin temperature if it gets cold?

.... *(1 mark)*

Guided **(b)** As your body temperature starts to drop, you may start to shiver. Explain why shivering is important to the body.

Shivering is the process where .... in the body ....

.... This process helps, because it .... *(3 marks)*

(c) Describe how blood vessels in the skin can also help to preserve the body's temperature.

....

....

.... *(3 marks)*

B-A* **3** The diagram shows a cross-section through human skin.

HIGHER

Explain how the structures in the skin help the body to regulate its temperature if the body becomes too hot.

The diagram here has only limited labels, so not all the parts of the skin that are used to help regulate temperature are labelled. Remember that your explanation should identify the parts of the skin that are involved.

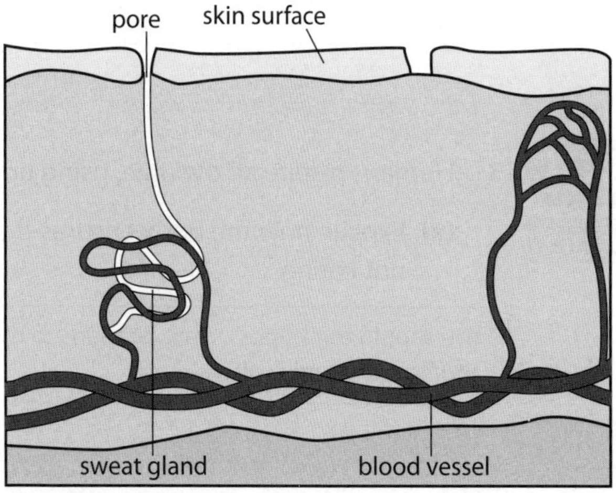

pore   skin surface

sweat gland   blood vessel

....

....

....

....

.... *(4 marks)*

Had a go ☐ Nearly there ☐ Nailed it! ☐

Blood glucose control

G-E **1** Draw a ring around the correct answer to complete the following sentences.

    (a) The level of glucose in the blood is measured in an organ called the | brain   heart   pancreas | .

(1 mark)

    (b) Blood glucose levels are controlled when this organ releases | an enzyme   a hormone   a sugar | .

(1 mark)

D-C **2** The graph shows how the concentration of glucose changes in a healthy person's blood during the day.

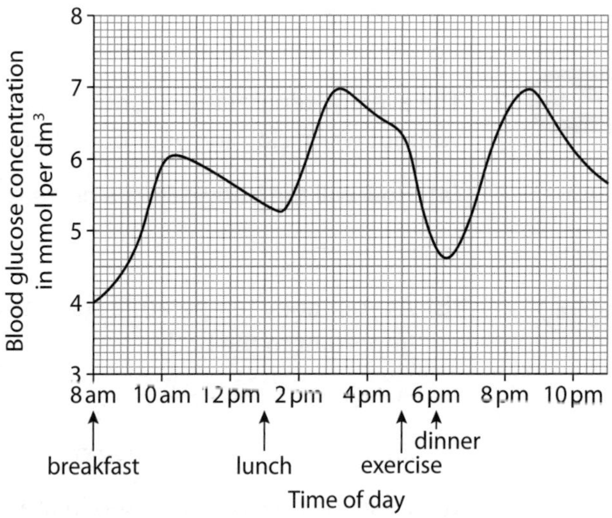

    (a) Name the substance, produced by the pancreas, that regulates blood glucose concentration.

        ....

(1 mark)

> **Guided** **(b)** Describe the trend in blood glucose concentration after a meal is eaten, using data from the graph to support your answer.

        The concentration of glucose in the blood .... after each meal.

        For example, after breakfast the change is from .... to ....

(2 marks)

    (c) Suggest how the graph would differ in a person with Type 1 diabetes.

        ....

        .... *(2 marks)*

B-A* **3** Humans respire all the time, using up glucose from the blood stream.

HIGHER

    (a) Explain how the body ensures that respiration can take place through the night when we are not eating.

      | You should think about what happens to the glucose concentration when we do not eat for some time, and how the body will respond. |

        ....

        ....

        .... *(3 marks)*

    (b) Control of the concentration of glucose in the blood is more difficult for people with diabetes. Explain why diabetics need to be very careful with their diet.

        ....

        ....

        .... *(2 marks)*

Biology six mark question 2

People who have kidney failure have options for treatment. These options include having a kidney transplant or using kidney dialysis.

Data produced by the NHS in England show that the survival rate for a kidney transplant, five years after the transplant, is between 80% and 86%, depending on whether the kidney was transplanted from a deceased donor or a live one.

Statistics for survival rates from dialysis are more difficult to obtain, as many people who start dialysis will then have a transplant. However, studies in America show that the survival rate, after five years, is between 75% and 85%, depending on the type of dialysis and whether it is carried out at home or in hospital.

Using this information, and your own knowledge, evaluate the different treatments available for kidney failure.

You will be more successful in six mark questions if you plan your answer before you start writing. There's no need to write about how these different treatments work – you should concentrate on the advantages and disadvantages of the two treatments.

Factors that you will need to think about are:

- What is the availability of the treatment?
- Does the treatment 'cure' the patient? Or is the benefit only short-term?
- What are the potential risks associated with the treatments?

You're given some data about success rates here – so make sure that you summarise these as part of your answer, too.

....

....

....

....

....

....

....

....

....

....

....

....

....

....

....

....

.... *(6 marks)*

Pollution

G-E **1** Draw a ring around the correct answer to complete the following sentence.

Increasing the amount of waste that we produce can cause

| deforestation |
| pollution |
| quarrying |

.

(1 mark)

D-C **2** Most of the time, the UK has westerly winds. These winds blow air from the UK towards Norway and Sweden. During the 1970s and 1980s, pine forests in Norway and Sweden were badly affected by acid rain. The governments of Norway and Sweden blamed the UK for polluting the air that was blown towards them.

Guided

(a) Describe how acid rain is formed.

The gas .... is released into the air, usually from burning fuels.

This gas then .... in rain water, turning it acidic.   *(2 marks)*

AQA SKILL
Analyse
Page 98

(b) The graph shows the level of acidity (pH) of a lake in Norway, measured over 15 years.

Use information from the graph to explain whether acid rain is still a problem in Norway.

First, you need to identify the trend or pattern in the data – this can be difficult with real data such as this. You can then work out whether acid rain is increasing or decreasing. Remember that lower pH values mean the rain is more acidic.

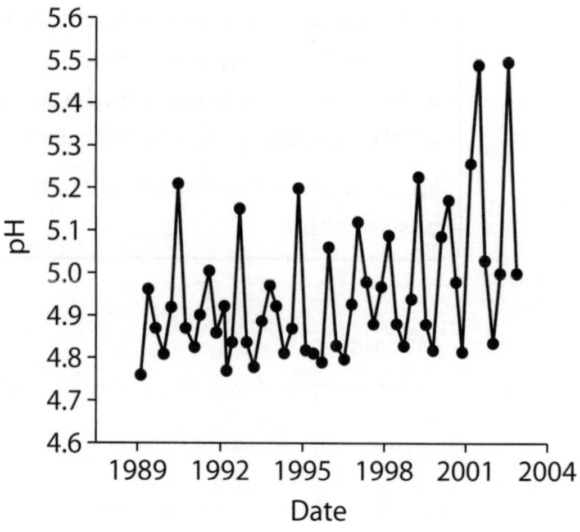

....

....

....   *(3 marks)*

B-A* **3** **(a)** A piece of woodland is converted to be used for growing crops. Explain how ground water coming from the piece of land may be affected if the farmer develops the land to get the largest yield of crops.

AQA SKILL
Describe
Page 98

....

....   *(2 marks)*

(b) Other than the effect on water supplies, converting land for farming can have other effects, especially on the environment. Considering the effects on the environment, describe the benefits and drawbacks of farming to produce our food.

....

....

....

....   *(4 marks)*

Deforestation

G-E 1 Rainforests in Indonesia are being cut down. The pie chart shows how the deforested land is used.

(a) What is the total proportion of land that is used for farming after deforestation?

....

....

(1 mark)

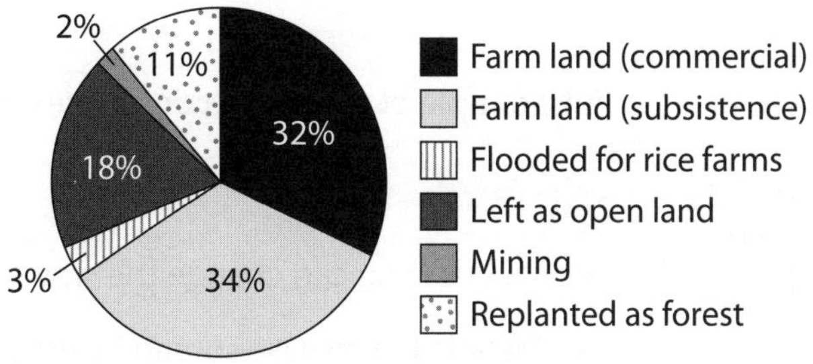

- ■ Farm land (commercial)
- ▨ Farm land (subsistence)
- ▥ Flooded for rice farms
- ■ Left as open land
- ▨ Mining
- ▨ Replanted as forest

2% 11% 32% 18% 3% 34%

(b) Give **one** use for the trees that are cut down in rainforests.

.... *(1 mark)*

Guided (c) 11% of the land that is cleared is replanted as forest. Why is replanting some trees important in terms of controlling carbon dioxide in the atmosphere?

Planting trees means that the rate at which carbon dioxide is taken from the atmosphere

.... The carbon dioxide is 'locked up' as ....

(2 marks)

D-C 2 The trees that are cut down during deforestation can be used to make wooden objects. Often, however, they are left to rot or burned.

EXAM ALERT

(a) Deforestation often leads to a loss of biodiversity.

(i) What is meant by the term **biodiversity**?

....

....

(1 mark)

> Be prepared to make connections such as the link between deforestation and an increase in carbon dioxide levels.

> Students have struggled with questions like this in recent exams – **be prepared!**

(ii) Explain why deforestation can lead to a loss of biodiversity.

....

.... *(2 marks)*

(b) Explain how leaving trees to rot has the same effect on the atmosphere as burning them.

....

....

.... *(3 marks)*

B-A* 3 Compost is used by many gardeners for growing plants from seeds. Some compost sold in the UK contains peat. Suggest the effect of using peat-based composts.

**AQA SKILL
Suggest
Page 98**

> Your answer needs to consider the effect that the removal of peat has on the environment.

....

....

....

.... *(4 marks)*

Global warming

 1 Draw a ring around the correct answer to complete the following sentence.

An example of a gas that contributes to global warming is

| carbon monoxide |
| methane |
| oxygen |

.

(1 mark)

 2 Global warming can change the way in which organisms behave.

> **Guided**

(a) Malaria is a disease spread by mosquitoes. The mosquitoes that carry the malaria parasite live in tropical areas of Asia and Africa. Explain how global warming could affect the spread of malaria.

If the temperature rose, there could be a change in ....

This would mean that malaria could spread ....

.... *(2 marks)*

(b) The Wildfowl and Wetlands Trust has reported fewer birds visiting Britain in the winter months. Explain how global warming could lead to this effect.

> There are two possible changes that could happen to behaviour of birds: one leading to an increase in birds in the UK in winter, and the other leading to a loss. Make sure you select the right one!

....

.... *(2 marks)*

B-A* **3** Carbon dioxide can be removed from the atmosphere by plants.

(a) Name the process that involves plants removing carbon dioxide from the atmosphere.

.... *(1 mark)*

(b) Describe another way in which carbon dioxide can be removed from the atmosphere.

.... *(2 marks)*

(c) The graph shows the carbon dioxide concentration in the atmosphere, and the change in atmospheric temperature. Evaluate the evidence the data gives about global warming.

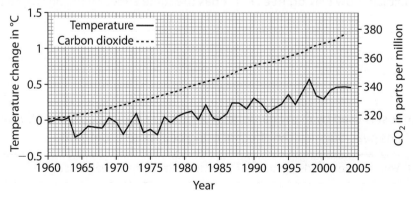

....

....

....

.... *(4 marks)*

Biofuels

G-E **1** Draw a ring around the correct answer to complete the following sentences.

(a) The production of biogas from waste
plant material involves the waste being [combusted   fermented   respired] . *(1 mark)*

(b) The main component
of the biogas produced is [carbon dioxide   hydrogen   methane] . *(1 mark)*

D-C **2** In a small village in rural Africa, a biogas generator produces fuel for the surrounding community.

(a) The biogas plant uses bacteria to produce the biogas. Name **two** sorts of material that the community can add to the biogas plant for the bacteria to digest.

....

.... *(2 marks)*

(b) Name the process in the bacteria that breaks down this material into biogas.

.... *(2 marks)*

Guided (c) Explain why small biogas fermenters like this can be useful in remote villages in developing countries.

The community may not have another ...., and – in developing

countries – might not be able to .... *(2 marks)*

B-A* **3** The South Shropshire biogas generator is based at Ludlow. Each year it consumes 5000 tonnes of garden waste and food waste that is separated from other household waste.

(a) Suggest why the plant needs to have food waste that is separated from other household waste.

....

.... *(2 marks)*

(b) The graph shows how the rate of biogas production changes with temperature in the generator.

(i) Describe how the temperature of the generator changes the rate of biogas production.

....

....

....

.... *(2 marks)*

(ii) Suggest how the generator at Ludlow is designed to overcome any issues raised by the effect of temperature on biogas production.

> Your answer should consider what the problems might be (think about the climate in the UK!), and then how the building of the generator helps address them.

....

....

.... *(3 marks)*

Food production

 1 The diagram shows the energy transfers occurring when a cow eats crops.

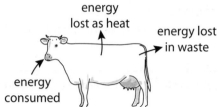

energy lost as heat

energy lost in waste

energy consumed

(a) The cow loses 1020 kJ in heat energy. Name the process happening in the cow that generates this heat energy.

.... *(1 mark)*

(b) The cow takes in 3050 kJ of energy from the crops it eats and loses 1910 kJ in waste.

(i) Name **one** waste material produced by the cow.

.... *(1 mark)*

 (ii) Calculate the amount of energy stored in the tissues of the cow.

Energy stored in tissues of cow = ... − .... − .... = ... kJ

(2 marks)

 2 Use your answer to the calculation in question **1 (b) (ii)** to explain why many people think it is better to produce crops for food, rather than producing meat to eat.

....

....

.... *(3 marks)*

B-A* **3** A student investigated the efficiency of farming chickens indoors and outdoors. He took two beakers with the same volume of hot water to represent chickens and placed one on the bench and one inside a cardboard box. He measured the temperature of the water over a 30-minute period, and repeated the investigation three times. His results are shown on the graph.

AQA SKILL
Evaluate
Page 98

Use these results, and your own knowledge, to evaluate the farming of chickens indoors compared to outdoors.

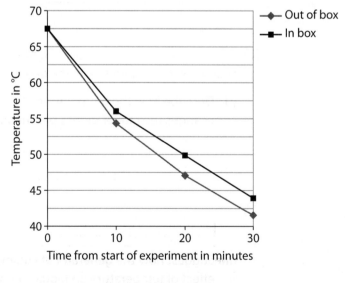

....

....

....

....

....

.... *(6 marks)*

Fishing

G-E 1 British fishermen have strict rules on the numbers and types of fish that they are allowed to catch in the North Sea.

(a) Describe how the quota system helps manage fish stocks.

....

.... *(2 marks)*

(b) Other countries also take fish from the North Sea. How might this affect the success of the quota used by British fishermen?

....

.... *(2 marks)*

D-C 2 The pie charts show the proportions of salmon used for human consumption that are caught in the wild or are farmed. The data are shown for the years 1982 and 2007.

1982   2007

■ Wild
☐ Farmed

(a) Describe what the pie charts show, using data to support your answer.

....

....

.... *(2 marks)*

> When you are asked to describe a difference between two charts, try to include some numbers in your answer.

Guided (b) Farmed salmon are kept in large numbers in cages in the sea. Salmon are mostly carnivorous and are fed on pellets of fish food. Farmed salmon are often higher in fat content than wild salmon. Use the information above to suggest why.

Farmed salmon are kept in cages, so ....

As they are provided with food, ... *(2 marks)*

(c) As humans eat more farmed salmon, there is an effect on wild salmon and other fish. Explain how farming salmon could alter the stocks of wild salmon and of other fish.

....

....

.... *(3 marks)*

B-A* 3 Explain how the dimensions of a fishing net can contribute to the conservation of fish stocks.

> Your answer here should think about how the right sort of net can select the type of fish caught, and the differences between those fish that are caught and those that are not.

....

....

....

.... *(4 marks)*

Sustainable food

G–E   **1**   A fermenter is used to grow the fungus *Fusarium*, which is used to produce mycoprotein.

    (a) What is the food source for the *Fusarium* in the fermenter?

    .... *(1 mark)*

> **Guided**

    (b) Explain the purpose of the air being bubbled into the base of the fermenter.

      The air contains ....,

      which the fungi need to ....

      .... *(2 marks)*

Diagram labels: Food in, Vent, Water out, Water jacket, Paddle, Water in, Air in

D–C   **2**   The table shows how 100 g of Quorn, a mycoprotein, compares with 100 g of other foods in terms of nutrient content.

	Quorn	Chicken	Beef	Fish	Cheese
Total fat in g	2.9	4.5	18.2	0.9	33
Total carbohydrate in g	9	0	0	0	1.3
Protein in g	11.5	30.9	26	23	25

    (a) Suggest the substance that makes up most of the rest of the mass of these foods.

    .... *(1 mark)*

    (b) Mycoprotein has the potential to solve issues of food supply that meats cannot. Explain why.

    > Your answer should consider the availability of different foods and issues of energy use in production.

    ....

    .... *(2 marks)*

B–A*   **3**   Thirty years ago, shops in the UK would only sell vegetables that were 'in season' and grown in the UK. Now, supermarkets import vegetables to sell all year around. Many imports come from Israel, where desert land has been irrigated to make farming land. Evaluate the advantages and drawbacks of this situation.

AQA SKILL Evaluate Page 98

    ....

    ....

    ....

    .... *(4 marks)*

B–A*   **4**   The fungus that grows in a fermenter is a living organism. Suggest why it is important to have a cold water jacket around the fermenter.

    ....

    ....

    .... *(3 marks)*

Biology six mark question 3

The graph shows how the area of land being deforested in Brazil changed between 1991 and 2009.

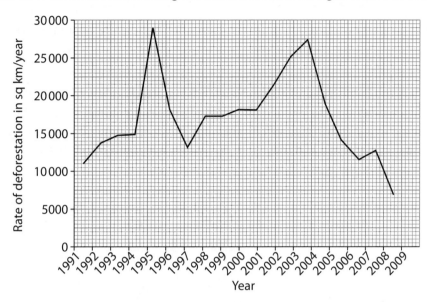

It is estimated that about half the world's tropical forests have been cleared for farm land and for timber. The remaining forests are estimated to store around 280 billion tons of carbon in their biomass.

Explain how deforestation can contribute to climate change.

> You will be more successful in six mark questions if you plan your answer before you start writing.
>
> Use data from the graph to show a trend.
>
> Your answer should also consider how aspects of deforestation may further contribute to climate change. What is the land that is cleared used for? What is the wood cleared from the land used for? The question has given you an extra piece of data after the graph, so do use this to help build a better answer.

....

....

....

....

....

....

....

....

....

....

....

....

....

....

.... *(6 marks)*

The early periodic table

The periodic table on page 95 may help you to answer these questions.

G-E

1 What property was used to order the elements in the early versions of the periodic table?

.... *(1 mark)*

D-C

2 In 1864 the English chemist John Newlands reported his 'law of octaves'. This stated that if the known elements were listed in the order of increasing atomic weight, then certain properties recurred every eighth element. There were, however, some obvious failings in his ideas on classifying elements, and they were not widely accepted by other scientists at the time.

A copy of his periodic table is shown.

Guided

(a) In general terms, how are the groups of elements with similar properties arranged differently in Newlands's table compared with the modern periodic table?

H	1	F	8	Cl	15	Co/Ni	22	Br	29	Pd	36	I	42	Pt/Ir	50
Li	2	Na	9	K	16	Cu	23	Rb	30	Ag	37	Cs	44	Tl	53
Gi	3	Mg	10	Ca	17	Zn	25	Sr	31	Cd	34	Ba/V	45	Pb	54
Bo	4	Al	11	Cr	18	Y	24	Ce/La	33	U	40	Ta	46	Th	56
C	5	Si	12	Ti	19	In	26	Zr	32	Sn	39	W	47	Hg	52
N	6	P	13	Mn	20	As	27	Di/Mo	34	Sb	41	Nb	48	Bi	55
O	7	S	14	Fe	21	Se	28	Ro/Ru	35	Te	43	Au	49	Os	51

In Newlands's table the groups are arranged in horizontal rows.

.... *(2 marks)*

(b) Name the whole group that is missing from Newlands's table.   .... *(1 mark)*

(c) Some of the symbols for the elements have been changed. What are the modern symbols for elements number 3 and 4 in Newlands's table?

.... *(1 mark)*

Guided

(d) Using the group containing the alkali metals, describe where Newlands's law of octaves works and where it fails.

In the group starting with lithium the law of octaves works up to ....

....

.... *(3 marks)*

The term 'periodic' in periodic table refers to the properties of elements that recur at regular intervals, and so elements are placed in groups with similar properties.

B-A*

3 Dmitri Mendeleev published his version of the periodic table in 1869.

(a) In what order did Mendeleev place the elements in his periodic table?

.... *(1 mark)*

(b) Mendeleev put question marks in for the elements with atomic weights 68 and 70.

(i) What are the modern names for these two elements?

.... *(1 mark)*

(ii) Give a reason why the symbols for these elements were missed out in his periodic table.

.... *(1 mark)*

The modern table

The periodic table on page 95 may help you to answer these questions.

D-C 1 The diagram below shows the position of six different elements in the periodic table. The letters do not represent the symbols for the elements.

	A																
																	B
															E	C	D
	F																

   (a) Identify **two** metals with the same number of electrons in their outer shell. .... *(1 mark)*

   (b) Identify the element that has one more electron in its highest occupied energy level than the element C. .... *(1 mark)*

   (c) Identify **two** elements that have the same number of occupied energy levels. .... *(1 mark)*

B-A* 2 The organisation of the modern periodic table is based on the atomic structure of the elements.

   (a) Explain how the elements are arranged in the modern periodic table in terms of their number of protons and electronic structure.

> Guided

   In the modern periodic table elements are arranged in order of ....

   number of protons, with elements placed in the same group if ....

   ....

   .... *(3 marks)*

   (b) (i) What is the general link between the main group number in the periodic table and an element's electronic structure?

   ....

   .... *(1 mark)*

   (ii) What is the exception to this rule?

   .... *(1 mark)*

B-A* 3 Most scientists did not think the early tables of elements were very useful.

   (a) Describe **two** changes Mendeleev made, when constructing his periodic table, that helped overcome some of the problems of elements that didn't fit correctly.

   ....

   .... *(2 marks)*

   (b) Suggest **one** possible reason why scientists did not appreciate the importance of the early attempts at classifying elements in a periodic table. The reason should be in addition to the answers to question **4 (a)**.

   .... *(1 mark)*

   (c) Describe **two** ways we can use the modern periodic table to make predictions about the properties of certain elements.

   ....

   .... *(2 marks)*

Group 1

> The periodic table on page 95 may help you to answer these questions.

G-E 1 Choose your answer to these questions from the following alkali metals.

lithium	potassium	rubidium	sodium

   (a) Which of these metals has the lowest melting point? .... *(1 mark)*

   (b) Which of these metals is the most reactive? .... *(1 mark)*

2 **(a)** Describe the bonding type, solubility and general colour of most compounds formed between alkali metals and non-metals.

   ....

   ....

   .... *(3 marks)*

> **Guided**

   (b) Explain why Group 1 elements are called the alkali metals. Include a chemical reaction as part of your explanation.

   When all alkali metals react with water they produce ....

   which dissolve in water to form .... *(3 marks)*

B-A* 3 Sodium is a typical alkali metal. The word equations for some reactions of sodium are shown below.

   sodium + chlorine → sodium chloride

   sodium + oxygen → sodium oxide

   sodium + water → .... + ...

   (a) Complete the word equation for the reaction of sodium with water. *(2 marks)*

   (b) Write balanced symbol equations for the reactions of sodium with chlorine and sodium with oxygen.

   (i) .... + ... → .... *(2 marks)*

   (ii) .... + ... → .... *(2 marks)*

   (c) Why is sodium stored under oil?

   ....

   .... *(2 marks)*

B-A* 4 A lab technician wanted to dispose of an old tin full of sodium metal. He decide he would get rid of it by rowing out to sea, punching holes in the tin and dropping it in the sea.

   State **two** facts about sodium and use them to explain why this is not a safe way of disposing of the metal.

   ....

   ....

   .... *(4 marks)*

Transition metals

> The periodic table on page 95 may help you to answer these questions.

1 Tick (✓) true or false for each of the following general statements comparing the properties of transition metals with alkali metals.

Statements	True	False
Alkali metals have higher densities than transition metals		
Transition metals are stronger than alkali metals		
Alkali metals are harder than transition metals		
Transition metals are more reactive than alkali metals		

(4 marks)

2 This question refers to the elements in the different blocks of the periodic table shown on the right.

Which of the blocks A to E contain:

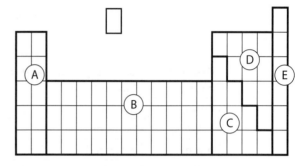

(a) mostly non-metals?

.... *(1 mark)*

(b) the group of elements that usually form ions with a 1+ charge? .... *(1 mark)*

(c) metals that are often used as catalysts? .... *(1 mark)*

(d) the most unreactive elements? .... *(1 mark)*

3 (a) Describe where the transition elements are found in the first 54 elements of the periodic table.

The transition elements are found between calcium and ....

and between .... and .... *(2 marks)*

(b) State how the melting points of alkali metals generally compare with the melting points of transition metals.

....

.... *(1 mark)*

4 Some metals can form more than one compound when combining with another element. For example, iron can form two different compounds with halogens like chlorine; these are iron(II) chloride and iron(III) chloride.

(a) Write down the formula for each of these iron compounds.

.... *(2 marks)*

(b) Write down the formula for the **one** compound formed between sodium and chlorine.

.... *(1 mark)*

(c) Explain why elements like iron can form more than one compound with the same element while alkali metals like sodium only form the one compound.

....

....

.... *(2 marks)*

Group 7

> The periodic table on page 95 may help you to answer these questions.

D-C **1** Describe the trend in reactivity, melting point and boiling point down the Group 7 elements.

As you go down the group, the reactivity ....

Guided

and melting and boiling point .... *(2 marks)*

B-A* **2** An experiment involving the reactions of Group 7 elements with iron is shown in the diagram.

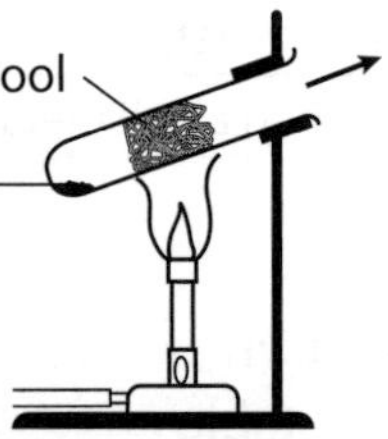

iron wool

a few crystals of iodine or a few drops of bromine

fumes released into fume cupboard

Group 7 element	Description of reaction with fine iron wool	Name and formula of main compound formed
fluorine	not carried out	–
chlorine	burns brightly	.... $FeCl_3$
bromine	glows red	.... $FeBr_2$
iodine	iron changes colour slowly	Iron(II) iodide, FeI_2

(a) (i) Why is the experiment carried out in a fume cupboard?

.... *(1 mark)*

(ii) Suggest **one** reason for not carrying out the reaction with fluorine.

.... *(1 mark)*

(b) Name the compounds of chlorine and bromine that are missing from the table.

.... *(2 marks)*

(c) Write a balanced chemical equation for the reaction of iron wool with iodine to form FeI_2.

.... *(2 marks)*

HIGHER **(d)** Explain in terms of electronic structure why the reactivity of the Group 7 elements decreases down the group.

....

....

.... *(3 marks)*

B-A* **3** When chlorine is bubbled through a solution of sodium bromide, bromine is formed in solution.

(a) Complete the balanced symbol equation for this reaction.

$Cl_2(g) + 2NaBr(aq) \rightarrow$ .... (aq) + ... (aq)   *(2 marks)*

(b) What is this kind of reaction called? .... *(1 mark)*

(c) State the rule that describes which halogens and halide ions will react together.

....

.... *(1 mark)*

Hard and soft water

G-E **1** Name **two** metal ions which are commonly found in hard water.

.... *(2 marks)*

D-C **2** What is the difference between hard and soft water in terms of:

(a) the action of soap in the water.

⟩ Guided ⟩

Soap forms a lather in soft water but forms a ....

.... *(2 marks)*

(b) the dissolved substances they contain.

....

.... *(2 marks)*

B-A* **3** A group of students wanted to compare the hardness of three different samples of water.
They were given the following apparatus and chemicals.

| **3 samples of water**   **100 cm³ conical flask**   **burette** |
| **measuring cylinder**   **soap solution** |

(a) Describe how the students could use this apparatus to compare the hardness of the three water samples.

....

....

....

.... *(4 marks)*

(b) To make the comparisons between the water samples fair the other variables in the test must be kept the same. State **two** variables that the students would need to keep the same for a fair test.

Calcium and magnesium ions form a precipitate if carbonate ions are added. The precipitates formed are solid calcium carbonate and magnesium carbonate.

....

.... *(2 marks)*

B-A* **4** Temporary hard water, which contains hydrogencarbonate ions (HCO_3^-), is found in certain areas.
This kind of hard water can be softened by boiling the water.

HIGHER

(a) Where do the ions in hard water come from?

.... *(1 mark)*

(b) Explain how boiling temporary hard water can remove the ions that make the water hard.

....

....

....

.... *(3 marks)*

Softening hard water

G-E   **1**   What is removed from hard water when it is softened? .... *(1 mark)*

D-C   **2**   A hard water supply can cause problems in the home in different ways.

- The water can have an odd taste.

- Soap doesn't work properly.

- A scale can form on the inside of kettles.

But some industries, like brewing, prefer a hard water supply.

(a) Why does soap not work well in hard water areas?

....

....

.... *(3 marks)*

(b) What is the scale that forms on the inside of kettles in hard water areas?

.... *(1 mark)*

> **Guided**

(c) Describe **two** possible health benefits of having a hard water supply.

The calcium in hard water helps ....

It also helps reduce .... *(2 marks)*

B-A*   **3**   Permanent hard water contains compounds like calcium chloride ($CaCl_2$) and magnesium sulfate ($MgSO_4$), and cannot be softened by boiling. Permanent hard water can be softened by adding sodium carbonate solution or by passing the water through an ion exchange resin.

(a) Why can permanent hard water **not** be softened by boiling?

....

.... *(2 marks)*

(b) The balanced equation below shows what happens when sodium carbonate solution is added to calcium chloride solution.

$CaCl_2(aq) + Na_2CO_3(aq) \rightarrow CaCO_3(s) + 2NaCl(aq)$

(i) Explain how this reaction softens the water.

....

....

.... *(3 marks)*

(ii) Write a similar balanced equation, with state symbols, for the reaction between magnesium sulfate solution and sodium carbonate solution.

.... *(2 marks)*

(c) What happens in an ion exchange resin to soften hard water?

....

.... *(2 marks)*

Purifying water

1 The main stages in water treatment are shown in the diagram.

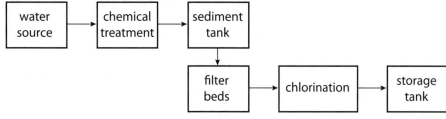

(a) Suggest **one** problem that might make the quality of the water from a particular source, such as a river, unsuitable for water supply.

.... *(1 mark)*

(b) Part of the chemical treatment involves adding aluminium sulfate as a coagulant to make small particles stick together to form larger lumps.

    (i) Write the chemical formula for aluminium sulfate.   .... *(1 mark)*

    (ii) Suggest what happens in the sediment tank.

       In the sediment tank the lumps of solid .... *(1 mark)*

(c) What happens in the filter beds?

.... *(1 mark)*

(d) Explain why chlorination is necessary.

....

.... *(2 marks)*

(e) Fluoride is often also added to our water. Describe **one** advantage and **one** disadvantage of adding fluoride to our water supply.

....

.... *(2 marks)*

2 Distillation can be used to produce pure water from almost any source.

> Distillation involves evaporating a liquid by heating it to its boiling point. This is followed by cooling to condense the gas formed back to a liquid, which is collected in a separate container.

(a) Explain how distillation produces pure water.

....

.... *(2 marks)*

(b) Why is distillation not used in hard water areas to improve the quality of the drinking water?

....

.... *(2 marks)*

(c) Suggest what kind of substances would not be removed from water by distillation.

....

.... *(1 mark)*

Chemistry six mark question 1

The alkali metals, Group 1, and the halogens, Group 7, are at opposite ends of the periodic table.

Explain the trends in the chemical reactions of Group 1 and Group 7 and elements.

Li														F	
Na														Cl	
K														Br	
Rb														I	
Cs														At	

> You are more likely to get better marks in six mark questions if you plan your answer before you start writing.
>
> Note that the question asks for trends in the reactions, not the properties, so do not waste time writing about properties. You need to describe the reactions to explain the trends. You should also give balanced equations for the reactions if you can.
>
> Make sure that you organise your answer clearly. You can divide the answer by groups or by chemical reaction.

....

....

....

....

....

....

....

....

....

....

....

....

....

....

....

....

....

....

....

.... *(6 marks)*

Calorimetry

1 A calorimeter can be used to measure the relative amount of heat energy released when a fuel burns. An example of a simple calorimeter is shown.

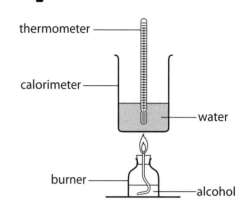

A student was asked to compare the heat energy output per gram for two different fuels called methanol and ethanol.

Complete the list of the **four** measurements that the student should take so that they can compare the heat energy output from the two fuels.

Mass of burner + fuel at start, ...

Temperature at start, ... *(3 marks)*

2 **(a) (i)** What would be the source of the greatest error in the experiment in question **1**?

.... *(1 mark)*

(ii) How could this error be reduced?

.... *(2 marks)*

(b) Explain how the results of the experiment in question **1** could be used to compare the heat energy released by the fuels.

....

.... *(2 marks)*

3 A group of students investigated the heat energy released when solid sodium hydroxide was dissolved in water.

The results of their experiments are shown in the table. 100 cm³ of water was used each time.

Measurement	Result
Mass of sodium hydroxide used in g	1.0
Temperature of water at start in °C	20
Temperature of water at end in °C	22

(a) Use the results of the experiment to calculate the energy released when 1.0 g of sodium hydroxide is dissolved in water. The equation that you need to use is: $Q = C \times m \times \Delta T$, where Q = energy released in joules, C = 4.2 (the heat capacity of water), m = the mass of water in grams and ΔT = change in temperature.

....

....

> Remember that 1 cm³ of water has a mass of 1 g (1 litre = 1 kg).

Energy released by 1 g = ... joules   *(3 marks)*

(b) Why should the students repeat the experiments?

.... *(1 mark)*

(c) What would the temperature rise be if the experiment was repeated using 2.0 g of sodium hydroxide instead of 1.0 g?

....

temperature rise = ... °C   *(1 mark)*

Energy level diagrams

1 The energy level diagram for the reaction A + B → C + D is shown in the diagram.

Which of the arrows 1 to 5 represent the following energy changes?

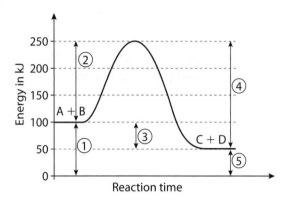

(a) The activation energy.

.... *(1 mark)*

(b) The overall energy change of the reaction.

.... *(1 mark)*

2 (a) Explain, using the energy level diagram in question **1**, if energy is given out or taken in during the reaction.

> **Guided**

The products have less energy than the reactants so energy has ....

.... *(2 marks)*

(b) What is the value of the overall energy change for the reaction in kJ?

.... *(1 mark)*

Most chemical reactions can be seen as a series of bond-breaking and bond-making steps. The activation energy, the energy required to start a reaction, is needed to break bonds. Energy is given out when bonds are formed.

B-A*

HIGHER

3 Use the energy diagram to answer these questions.

(a) What is the value, in kJ, for the overall energy change in this reaction?

energy change = ... kJ

(1 mark)

(b) What is the value, in kJ, for the activation energy for this reaction?

activation energy = ... kJ

(1 mark)

(c) With reference to bond making and bond breaking steps and the energy level diagram explain why this reaction is endothermic.

....

....

.... *(3 marks)*

(d) The presence of a catalyst would speed up this reaction.
Draw a line on the graph to represent the energy level diagram for this reaction when a catalyst is added.

(2 marks)

Bond energies

 This whole page is Higher material.

 1 Ammonia is made by the reaction of nitrogen and hydrogen as shown in the following balanced equation.

$N_2(g) + 3H_2(g) \rightarrow 2NH_3(g)$

The equation can be shown using structural formulae.

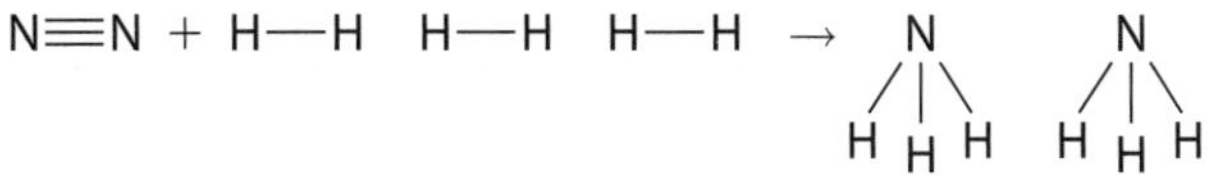

$N\equiv N$ + H—H   H—H   H—H $\rightarrow$ N   N
                        /|\ /|\
                       H H H H H H

Guided   **(a)** Use the bond energies in the table to calculate the energy change in this reaction.

Bond	Bond energy in kJ
N≡N	941
H—H	436
N—H	391

Energy in

1 × N≡N = 1 × 941 kJ = 941 kJ

3 × H—H = 3 × 436 kJ= ....

Total = ...

Energy out

6 × N—H =6 × ....

....

....

Energy change = ... kJ   *(3 marks)*

(b) Explain if the reaction is exothermic or endothermic in terms of bond energies.

....

.... *(2 marks)*

 2 The balanced equation for the combustion of methane is shown below.

$CH_4(g) + 2O_2(g) \rightarrow CO_2(g) + 2H_2O$

Explain why this reaction, although very exothermic, does not go spontaneously but needs an input of energy to start it off.

....

.... *(2 marks)*

3 The structural formulae equation for the reaction of hydrogen and bromine is shown below.

H—H + Br—Br → 2H—Br

Bond	Bond energy in kJ
H—H	436
Br—Br	193

(a) Use the bond energies in the table calculate the activation energy for the above reaction.

....

Activation energy = ... kJ   *(2 marks)*

(b) The reaction is exothermic and the overall energy change is 103 kJ for the reaction as written. Use this data and the information above to calculate the bond energy for the H—Br bond.

....

H—Br bond energy = ... kJ   *(2 marks)*

Hydrogen as a fuel

D-C 1 Many car manufacturers are investigating ways of using hydrogen gas as a fuel to power motorcars instead of using diesel and petrol.

Guided

(a) Describe **two** ways that hydrogen could be used as a fuel in motor cars.

Hydrogen can be burned as a fuel or it can be ....

.... *(2 marks)*

(b) Why is hydrogen thought to be a cleaner alternative fuel compared with diesel and petrol?

....

....

.... *(3 marks)*

B-A* 2 Hydrogen gas burns readily and can be used as a fuel. Two methods of making hydrogen gas are described below.

Method 1: the steam reforming of methane, which involves reacting water (H_2O) with methane (CH_4) in the presence of a catalyst. Carbon monoxide (CO) is produced as well as hydrogen (H_2).

Method 2: the electrolysis of water using electricity produced from solar panels or wind power. This also forms two gases, one of which is hydrogen (H_2).

(a) Write a balanced symbol equation for the combustion of hydrogen.

.... *(2 marks)*

(b) Write a balanced symbol equation for the production of hydrogen from methane and steam.

.... *(2 marks)*

(c) During the electrolysis of water, hydrogen is produced at the negative electrode.
What will be formed at the positive electrode?

.... *(1 mark)*

(d) Describe **two** advantages of using hydrogen produced by method 2 compared with method 1.

....

....

.... *(2 marks)*

(e) Explain why the electrolysis of water using electricity from the National Grid does not have the same advantages as the electrolysis described in method 2 above.

....

.... *(2 marks)*

B-A* 3 Hydrogen fuel cells are being developed as a source of energy for use in transport systems, industry and the home.

(a) Name the gas, other than hydrogen, used in a hydrogen fuel cell. .... *(1 mark)*

(b) Describe **two** problems that have to be overcome before we can use hydrogen fuel cells for cars.

....

.... *(2 marks)*

Tests for metal ions

G-E　**1**　Draw a ring around the correct answer to complete these sentences.

(a) A crimson flame is produced by
| sodium |
| lithium |
| barium |
.

(1 mark)

(b) Adding sodium hydroxide to a solution containing calcium ions forms a white
| solid |
| liquid |
| gas |
.

(1 mark)

D-C　**2**　A forensic scientist working at a crime scene tested an unknown solid found on a suspect's shoe. The results of their observations are shown below.

Appearance	Solubility	Test 1: Flame test	Test 2: Effect of adding dilute hydrochloric acid
white solid	insoluble	red flame	bubbles of gas formed that turns limewater cloudy

> You need to learn the flame colours produced by the following ions: lithium, sodium, potassium, calcium and barium.

(a) Describe how to carry out a flame test.

....

.... *(2 marks)*

Guided　**(b)** Why are at least two tests needed to identify any ionic substance?

　　Ionic substances contain two ions and .... *(2 marks)*

(c) Name the gas that is produced in test 2. .... *(1 mark)*

(d) Suggest a possible name for the unknown white solid. .... *(2 marks)*

B-A*　**3**　The effect of adding sodium hydroxide solution to solutions of different metal ions is shown in the table below.

Cation	Symbol	Effect of adding sodium hydroxide solution
aluminium	Al^{3+}(aq)	white solid formed
magnesium	Mg^{2+}(aq)	white solid formed
copper(II)	....	blue solid formed
iron(II)	Fe^{2+}(aq)	.... solid formed
iron(III)	Fe^{3+}(aq)	.... solid formed

(a) Name the type of reaction that occurs in these tests. .... *(1 mark)*

(b) Complete the missing information in the table. *(3 marks)*

(c) Describe how this test can be used to tell the difference between solutions containing aluminium and magnesium ions.

....

.... *(2 marks)*

(d) Complete the balanced equation, with state symbols, for the reaction that occurs when sodium hydroxide solution is added to calcium nitrate solution

　　2NaOH(aq) + Ca(NO$_3$)$_2$(aq) → .... (......) + 2NaNO$_3$ (......) *(3 marks)*

More tests for ions

D-C 1 A group of students were given a white soluble solid that was thought to be either potassium carbonate or potassium sulfate.

> **Guided**

Describe how the students could test the white solid to see if it contained carbonate ions or sulfate ions.

First add hydrochloric acid. If carbonate ions are present a gas is produced that will turn

....

To test for sulfate ions add ....

.... *(2 marks)*

B-A* 2 Describe how to carry out **one** test for each pair of substances to identify which is which. Use chemicals from the list in the box.

barium chloride solid	distilled water	hydrochloric acid solution
nitric acid solution	silver nitrate solid	sodium hydroxide solution

(a) Sodium bromide and sodium chloride.

....

.... *(2 marks)*

(b) Sodium sulfate and sodium chloride.

....

.... *(2 marks)*

B-A* 3 A sample of river water, taken downstream from a large town, was tested for pollution by analytical chemists working for the water authority.

AQA SKILL
Interpret
Page 98

The water sample was first evaporated until 5% was left and then tested as follows.

Test 1: Flame test results	Test 2: Adding sodium hydroxide solution	Test 3: Adding dilute hydrochloric acid	Test 4: Adding acidified silver nitrate solution
no colour produced	A white precipitate formed, which dissolved when excess hydroxide was added.	no effect	yellow solid formed

(a) Suggest what tests 1 and 2 would tell the analytical chemist about the river water.

....

.... *(2 marks)*

(b) Suggest what tests 3 and 4 would tell the analytical chemist about the river water.

....

.... *(2 marks)*

(c) There were several different factories sited on the banks of the river in the town. Suggest how you could carry out further tests to find out which factory was producing the pollution?

....

.... *(2 marks)*

Titration

Use the relative atomic masses on the periodic table on page 95 to help you answer these questions.

 1 Complete the following sentence using words from the box below.

test	hydroxide	indicator	neutraliser	precipitation	titration

The volumes of acid and alkali solution that exactly react with each other can be found by

carrying out a .... using a suitable .... to show

when the reaction is complete.   *(2 marks)*

 2 5.0 cm³ of nitric acid was needed to neutralise exactly 15.0 cm³ of 0.2 mol/dm³ lithium hydroxide solution. The balanced equation for the reaction is shown below.

$HNO_3(aq) + LiOH(aq) \rightarrow LiNO_3(aq) + H_2O(l)$

The number of moles in a solution can be calculated using the equation $N = C \times V$, where N = number of moles, C = concentration in mol/dm³ and V = volume in dm³.

 (a) Calculate the concentration of the nitric acid in mol/dm³.

volume = 15 cm³ ÷ 1000 = 0.015 dm³

Moles of LiOH = 0.015 × 0.2 = 0.003

1 mole of LiOH neutralises 1 mole of HNO₃ so .... moles of HNO₃ are present

$N = C \times V$, ... = $C \times$ .... so C = .../.............. = ...

concentration of nitric acid = ... mol/dm³   *(3 marks)*

(b) Calculate the concentration of the 0.2 mol/dm³ of lithium hydroxide in g/dm³.

Note: To convert cm³ to dm³ divide by 1000.

....

....   *(2 marks)*

 3 25.0 cm³ of sodium hydroxide solution is exactly neutralised by 7.5 cm³ of 1.0 mol/dm³ hydrochloric acid. The balanced equation for the reaction is shown below.

$HCl(aq) + NaOH(aq) \rightarrow NaCl(aq) + H_2O(l)$

(a) Calculate the concentration of the sodium hydroxide in mol/dm³.

....

....

....   *(3 marks)*

(b) Calculate the concentration of the 1.0 mol/dm³ of hydrochloric in g/dm³.

....

....   *(2 marks)*

Chemistry six mark question 2

In this question you will be assessed on using good English, organising information clearly and using specialist terms where appropriate.

The water supplied to our homes and factories can be classified into three types:

- soft water
- permanent hard water
- temporary hard water.

Describe the chemical and physical differences between these three types of water, including a description of the advantages and disadvantages of each type.

> You are more likely to get better marks in six mark questions if you plan your answer before you start writing. Your answer should include the following:
> - The chemical difference between soft water, permanent hard water and temporary hard water.
> - The properties of each type of water, and the advantages and disadvantages of each.

....

....

....

....

....

....

....

....

....

....

....

....

....

....

....

....

....

....

....

....

....

....

.... *(6 marks)*

The Haber process

G–E

1 Ammonia is manufactured industrially by the Haber process.

(a) Where are the nitrogen and hydrogen obtained from for the Haber process?

....

.... *(2 marks)*

> **Guided**

(b) What other substance is present in the reaction chamber and why is it added?

Iron is present and it acts as .... *(2 marks)*

D–C

2 (a) Why is it impossible to get **all** of the nitrogen and hydrogen converted into ammonia in the reactor during the Haber process?

.... *(1 mark)*

(b) The Haber process only produces about a 5% conversion of reactants. What is done to make sure the unreacted nitrogen and hydrogen is not wasted?

.... *(1 mark)*

B–A*

HIGHER

AQA SKILL Interpret Page 98

3 In the Haber process the reaction forming ammonia from nitrogen and hydrogen can be written as shown below.

nitrogen + hydrogen $\rightleftharpoons$ ammonia

(a) Complete the balanced equation below, including state symbols.

$N_2(g)$ + ... (......) $\rightleftharpoons$ .... (......)   *(2 marks)*

(b) What does the $\rightleftharpoons$ sign tell you about the reaction?

.... *(1 mark)*

B–A*

4 The graph shows how the yield of ammonia, in the Haber process, is affected by changes to temperature and pressure.

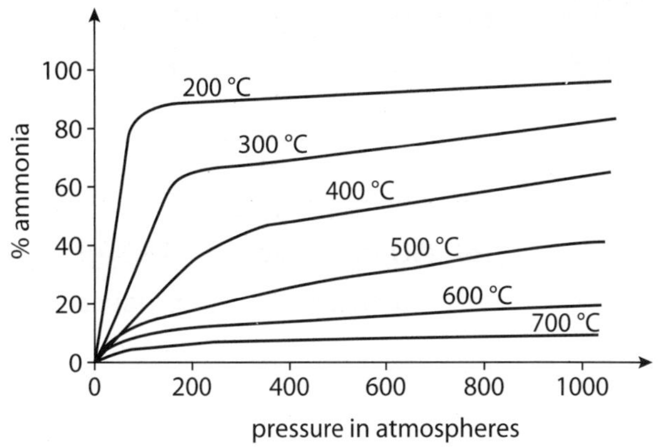

(a) The Haber process is normally carried out at 200 atmospheres pressure. Suggest **one** advantage and **one** disadvantage of increasing the pressure in the Haber process beyond 200 atmospheres pressure.

....

.... *(2 marks)*

(b) The Haber process is normally carried out at a temperature of 450 °C.

(i) With reference to the graph, give a reason why a lower temperature of 200 °C might be an advantage.

.... *(1 mark)*

(ii) Suggest a reason why a very low temperature is not used in this process.

.... *(1 mark)*

Equilibrium

HIGHER This whole page is Higher material.

 B-A*   **1**   Dinitrogen tetroxide (N_2O_4) is a pale yellow gas. It decomposes in an endothermic reaction to form the brown gas nitrogen dioxide (NO_2). The equation for the reversible reaction is hown below.

$$N_2O_4(g) \rightleftharpoons 2NO_2(g)$$
pale yellow   brown

(a) State what is meant by:

    (i) an **endothermic reaction** .... *(1 mark)*

    (ii) a **reversible reaction**. .... *(1 mark)*

Guided   **(b)** Explain what would happen to the colour of these gases if the temperature is raised from 20 °C to 40 °C.

     If the temperature is raised, the yield in the endothermic direction will increase, so there

     will be more NO_2, so the colour will ....

.... *(3 marks)*

(c) Explain what would happen to the colour of these gases if the gas pressure is increased.

....

....

.... *(3 marks)*

 B-A*   **2**   In each of the following reversible reactions explain how the amount of product formed at equilibrium is affected by decreasing the gas pressure.

(a) $H_2(g) + Br_2(g) \rightleftharpoons 2HBr_2(g)$
     reactants   products

....

.... *(2 marks)*

(b) $2SO_2(g) + O_2(g) \rightleftharpoons 2SO_3(g)$
     reactants   products

....

.... *(2 marks)*

 B-A*   **3**   Methanol is formed in industry by the exothermic reaction between carbon monoxide and hydrogen. The balanced equation for this reversible reaction is shown below.

$CO(g) + 2H_2(g) \rightleftharpoons CH_3OH(g)$

Why would the yield of methanol in this reaction be increased by lowering the temperature and raising the gas pressure?

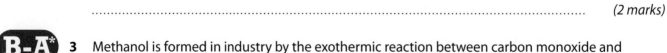

....

.... *(2 marks)*

Alcohols

1 The fourth member of the alcohol series is called butanol. Its structure is shown below.

H H H H
H—C—C—C—C—O—H
H H H H

(a) What is the functional group in the butanol molecule? .... *(1 mark)*

(b) What is the molecular formula of butanol? .... *(1 mark)*

> A functional group in a molecule is the group of atoms that gives it distinctive properties. For example, the functional group in alkenes is the C=C double bond.

2 **(a)** What **two** products would be formed by the complete combustion of butanol?

.... *(2 marks)*

(b) Suggest **two** possible uses for butanol.

Butanol could be used as a solvent or as a .... *(2 marks)*

3 Methanol and propanol are two members of the alcohol series. They have similar molecular structures and similar chemical properties.

(a) Write down the molecular formula for methanol and the displayed structural formula for propanol.

.... *(2 marks)*

(b) Draw a ring around the correct answer to complete the following sentences about the properties of the alcohols methanol and propanol.

Both these alcohols react with

| magnesium |
| copper |
| sodium |

metal to produce

| oxygen |
| hydrogen |
| carbon dioxide |

gas.

(2 marks)

4 Ethanol can be made by the fermentation of sugars from plants. In some countries ethanol is used as a fuel for cars in place of petrol. Ethanol is also found in alcoholic drinks. When certain alcoholic drinks are left exposed to the air they can turn into vinegar.

(a) Explain why ethanol produced by fermentation can be described as a renewable fuel.

....

.... *(2 marks)*

(b) Complete the balanced equation for the complete combustion of ethanol.

$C_2H_5OH + 3O_2 \rightarrow$ .... + ... *(2 marks)*

(c) State **one** health problem and **one** social problem associated with drinking alcohol.

....

.... *(2 marks)*

(d) Name the kind of reaction that forms vinegar from ethanol. .... *(1 mark)*

(e) What is the correct chemical name for the acid in vinegar? .... *(1 mark)*

Carboxylic acids

 1 The structural formula of butanoic acid, the fourth member of a homologous series of carboxylic acids, is shown to the right.

(a) What is the molecular formula of butanoic acid? .... *(1 mark)*

(b) Draw a ring around the correct general formula for carboxylic acids.

   $C_nH_{2n}O$ $C_nH_{2n}O_2$ $C_{n}H_{2n+2}O_2$ *(1 mark)*

 2 (a) Name the alcohol that could be oxidised to make butanoic acid. .... *(1 mark)*

(b) The third member of the carboxylic acid series is called propanoic acid. Write down the molecular formula of propanoic acid and draw its displayed structural formula.

....

(2 marks)

(c) What gas is produced when carboxylic acids, like propanoic acid and butanoic acid, react with metal carbonates?

.... *(1 mark)*

> **Guided** (d) Describe another common reaction of carboxylic acids, naming the general reactant and product involved.

   Carboxylic acids react with alcohols, forming .... *(2 marks)*

 3 The table below shows some information about two acid solutions.

HIGHER

Acid	Hydrochloric acid	Ethanoic acid
Formula	HCl	....
Concentration	0.1 mol/dm³	0.1 mol/dm³
pH	1	....
Time taken to dissolve 1 g of CaCO₃	....	6 minutes

(a) Complete the missing information in the table. *(3 marks)*

(b) The equation below represents what happens when ethanoic acid dissolves in water.

   $CH_3COOH(aq) \rightleftharpoons CH_3COO^-(aq) + H^+(aq)$

Use this equation to describe the difference between a weak acid like ethanoic acid and a strong acid like hydrochloric acid.

....

.... *(2 marks)*

(c) Explain the difference in pH between 0.1 mol/dm³ hydrochloric acid and 0.1 mol/dm³ ethanoic acid.

....

....

.... *(3 marks)*

Esters

G-E **1** What **two** types of compound react together to make an ester?

.... *(2 marks)*

D-C **2** Ethyl ethanoate can be prepared in the laboratory by the following steps:

- Put $2\,cm^3$ of .... acid and $2\,cm^3$ of .... into a test tube.

- Carefully add a few drops of concentrated sulfuric acid.

- Place the test tube in a water bath of warm water and leave for 4 minutes.

- Pour the contents of the test tube into a beaker containing sodium carbonate solution.

(a) What **two** compounds should be added to make ethyl ethanoate?

.... *(2 marks)*

(b) Why is concentrated sulfuric acid added to the mixture?

.... *(1 mark)*

Guided **(c)** Suggest a reason why a water bath is used, rather than a Bunsen burner.

The reactants (and products) in this reaction can be flammable so a water bath is

.... *(2 marks)*

(d) The final step of pouring the contents of the test tube into a beaker of sodium carbonate can be used to confirm the formation of an ester.

(i) What does the sodium carbonate react with in the mixture?

.... *(1 mark)*

(ii) Suggest **one** observation that would indicate that an ester had been formed.

....

.... *(1 mark)*

B-A* **3** The structural formula of an organic compound is shown here.

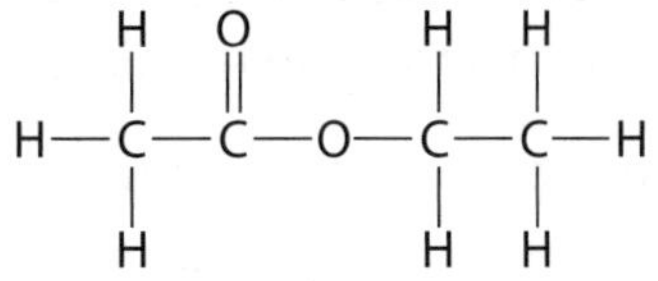

$$H-\underset{\underset{H}{|}}{\overset{\overset{H}{|}}{C}}-\overset{\overset{O}{||}}{C}-O-\underset{\underset{H}{|}}{\overset{\overset{H}{|}}{C}}-\underset{\underset{H}{|}}{\overset{\overset{H}{|}}{C}}-H$$

(a) Name this compound and state which homologous series it belongs to.

.... *(2 marks)*

(b) Draw the functional group in this compound.

.... *(1 mark)*

(c) Suggest **two** possible uses for this compound

.... *(2 marks)*

Using organic chemicals

G-E   **1** The structural formulae for three organic compounds are shown in the table.

Compound	Structural formula	Molecular formula	Type of compound
A	H O H | || | H—C—C—O—C—H | | H H	....	....
B	H | H H O H | | | | H—C—C—C—C—H | | | | H H H H	$C_4H_{10}O$	alcohol
C	H H H | | | O H—C—C—C—C⫽ | | | \ H H H O—H	$C_4H_8O_2$	carboxylic acid

(a) Complete the missing information in the table.   *(3 marks)*

(b) Which of the compounds, in the above table, is most likely to:

   (i) have a distinctive sweet smell?   ....   *(1 mark)*

   (ii) form a solution with a pH of 3?   ....   *(1 mark)*

D-C   **2** Indicate a use for each of the chemicals listed in the table below.
Tick (✓) **two** boxes for each chemical.

Chemical	Use			
	fuel	solvent	perfumes	food flavouring
ethyl ethanoate				
ethanol				

(2 marks)

B-A*   **3** The weak acid, ethanoic acid, can be made in different ways.

Method 1 involves passing natural gas, which is mainly methane, with air over a heated catalyst in a continuous process. The methane reacts with the oxygen in the air to form ethanoic acid.

Method 2 involves leaving alcohol solutions, produced by fermentation, exposed to the air for a few days. This allows time for microbes in the air to oxidise the ethanol to ethanoic acid.

(a) Complete the equations below, which represent each of these methods.

   Method 1: $2CH_4 +$ .... $\rightarrow CH_3COOH + 2H_2O$

   Method 2: .... $+ O_2 \rightarrow CH_3COOH + H_2O$ *(2 marks)*

(b) Give a reason why method 1 is likely to be the most economical method for the industrial manufacture of ethanoic acid in industry.

....

....   *(2 marks)*

(c) What is the main advantage of using ethanol, from fermentation, in method 2?

....

....   *(2 marks)*

Chemistry six mark question 3

Vinegar is a solution of ethanoic acid in water. Different types of vinegar contain different concentrations of acid.

You are given two bottles of different vinegars. Describe a method you could use to find out which vinegar contains the highest concentration of acid.

Your method should use all of the apparatus and chemicals listed below.

samples of vinegar X	50 cm³ measuring cylinder
samples of vinegar Y	200 cm³ beaker
dilute solution of sodium hydroxide	50 cm³ burette
indicator liquid	burette stand

You are more likely to get better marks in six mark questions if you plan your answer before you start writing. Your answer should include the following:
- How you will use the apparatus and chemicals in the investigation.
- What will be measured and how measurements will allow a conclusion to be made.
- How results can be made more repeatable, and any necessary safety precautions.

....

....

....

....

....

....

....

....

....

....

....

....

....

....

....

....

....

....

....

.... *(6 marks)*

X-rays

G-E 1 Which **one** of the following is a true statement about X-rays? Tick (✓) **one** box.

☐   A: X-rays are a form of electromagnetic radiation with a shorter wavelength than light.

☐   B: Prolonged exposure to X-rays is not harmful to health.

☐   C: X-rays are non-ionising radiation.

(1 mark)

G-E 2 Put ticks (✓) in the boxes in the table to show why an X-ray photograph shows bones but not the soft tissues.

	Transmit X-rays	Absorb X-rays
Bones and metals		
Soft tissues		

> Remember that X-ray photographs are usually 'negative' images.

(2 marks)

D-C 3 Describe **three** precautions that should be taken when X-rays are being used.

> **Guided**

X-ray sources should be shielded by lead to stop X-rays from being emitted in all directions.

....

....   *(3 marks)*

B-A* 4 X-rays were discovered in 1895. It soon became popular for people to view X-ray pictures of themselves showing their bones. Explain why X-rays of people are now only used for medical purposes.

....

....   *(2 marks)*

B-A* 5 The photo shows an X-ray of a human hand. The person is wearing a metal ring.

Explain what the image shows about the amount of X-rays transmitted and absorbed by different materials.

....

....

....

....

....

....

....   *(3 marks)*

Ultrasound

D-C 1 The echo of an ultrasound pulse from a foetus in a woman's womb is received by the detector 0.000 08 s after the probe sent out the pulse.

(a) Explain why there is an echo.

....

.... *(2 marks)*

(b) Calculate the distance to the foetus from the ultrasound probe. The speed of sound through the body is 1500 m/s. Write down the equation you use, and show clearly how you work out your answer. Give the unit of your answer.

> Ultrasound is often reflected – it goes there and back – so be careful in calculations about distance.

> In an exam, all the equations you need will be given to you on a separate sheet. In this book, the equations sheet is on page 97. At Higher Tier you are expected to be able to rearrange equations when necessary.

....

....

distance = ... unit .... *(3 marks)*

B-A* 2 Ultrasound can be used to examine the brains of babies because the parts of the skull are not fused together and do not block the ultrasound. A scientist uses an ultrasound probe connected to an oscilloscope. The probe is pressed against the head of a baby. The trace shows a peak when the sound is emitted from the probe and an echo from the skull on the opposite side of the brain.

scale = 5×10^{-6} s

Guided (a) Use information from the picture of the oscilloscope screen to calculate the time taken in seconds for the ultrasound pulse to pass through the skull.

time between peaks =

time taken for ultrasound to travel one way through skull =

....

time taken = ... s   *(2 marks)*

(b) In another test the ultrasound takes 8×10^{-6} s to travel across a baby's brain. The brain is 0.12 m in diameter. Calculate the speed of sound through brain tissue in metres per second. Write down the equations you use, and then show clearly how you work out your answer.

....

....

....

speed of sound = ... m/s   *(3 marks)*

Medical physics

G-E 1 The table shows some medical procedures. Tick (✓) the boxes in the table to show which type of wave is used for each procedure.

Medical method	Uses X-rays	Uses ultrasound
scanning pregnant women		
killing cancer cells in patients		
obtaining pictures of broken bones		
breaking up kidney stones		

(4 marks)

D-C 2 X-rays can cause cancers, while ultrasound is thought to be safe. State **one** difference in the properties of X-rays and ultrasound that explains this.

.... *(1 mark)*

D-C 3 Explain how images are produced in CT scans.

> **Guided**

In a CT scan, ... pass through the body. The source and detector, called a

...., rotate around the subject. An image is formed by combining

signals from .... *(3 marks)*

B-A* 4 Both X-ray CT scans and ultrasound are used for imaging the interior of human bodies.
Discuss the advantages and disadvantages of the use of X-rays and ultrasound in medical diagnosis.

AQA SKILL Compare Page 98

> Make sure that you know the advantages and disadvantages of CT scanners.

....

....

....

....

.... *(4 marks)*

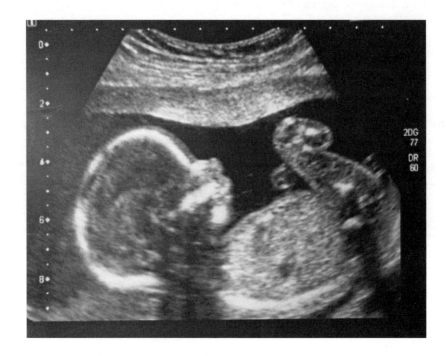

Refraction in lenses

  1 Draw a ring around the correct answer in the box to complete the following sentences.

Light changes direction passing from one medium to another. This is called $\boxed{\begin{array}{c}\text{reflection}\\\text{refraction}\\\text{focusing}\end{array}}$.

When light goes from air into glass it changes direction $\boxed{\begin{array}{c}\text{towards}\\\text{away from}\\\text{parallel to}\end{array}}$ the normal.

(2 marks)

Make sure you know which way light changes direction when it passes from one medium to another.

  2 The diagram shows light passing through a convex lens.

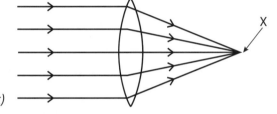

(a) Describe what happens to a parallel beam of light when it passes through a convex lens.

.... *(1 mark)*

(b) (i) What name is given to point X?

.... *(1 mark)*

(ii) What name is given to the distance from this point to the centre of the lens?

.... *(1 mark)*

  3 A ray of light is incident on a convex lens at an angle of 60°
and refracted at an angle of 33°. Calculate the refractive
index of the material from which the lens is made.
Write down the equation you use, and then show clearly
how you work out your answer.

> When calculating the refractive index, remember it is the sine of the angle, not the angle itself, that is needed.

$\text{refractive index} = \dfrac{\sin i}{\sin r} = \dfrac{\sin 60°}{\sin 33°} = $ ....

....

refractive index = ... *(3 marks)*

B-A* **4** A raindrop acts as a lens. The refractive index of water is 1.33. Light hitting a raindrop changes
direction to form an angle of 30° with the normal in the water. Calculate the angle of incidence.
Write down the equation you use, and then show clearly how you work out your answer.

....

....

....

....

angle of incidence = ... *(4 marks)*

Had a go ☐ Nearly there ☐ Nailed it! ☐

Images and ray diagrams

G–E **1** A digital projector contains a small display unit. Light from the display passes through a convex lens. It forms a large image on a screen. The object and the image are the opposite way up.

> **Guided**

Draw a ring around **three** words from the list below that describe the image.

(**inverted**) **magnified** **upright** **virtual** **real** **smaller**

(3 marks)

> Make sure that you know the difference between a real image and a virtual image.

D–C **2** Look at the ray diagram.

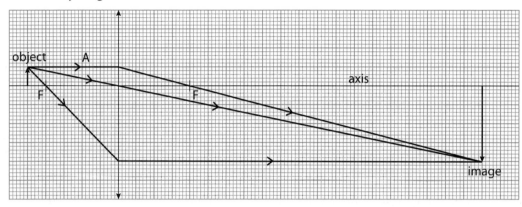

(a) Describe the path of the ray of light marked A as it travels from the object to the image.

....

.... *(2 marks)*

(b) Describe the image.

....

.... *(3 marks)*

(c) Describe the difference between a real image and a virtual image.

....

....

....

.... *(4 marks)*

B–A* **3** Complete the ray diagram below to show where the image is formed.

> **AQA SKILL** Draw Page 98

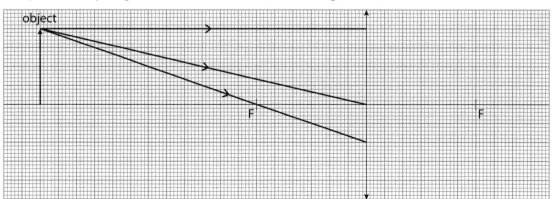

> Always use a ruler when drawing ray diagrams – and don't forget the arrows to show the direction of the rays.

(3 marks)

Real images in lenses

1 An object is 3.5 cm high. A lens produces an image of the object that is 70 cm high. What is the magnification of the image? Write down the equation you use, and then show clearly how you work out your answer.

$$\text{magnification} = \frac{\text{image height}}{\text{object height}} = \text{...}$$

magnification = ... *(2 marks)*

2 A lens has a focal length of 10 cm. Draw **one** line from each object distance to the correct image description.

Object distance	Image description

Object distance

- 20 cm
- 14 cm
- 22 cm

Image description

- real, inverted, same size
- real, inverted, smaller
- real, upright, magnified
- real, inverted, magnified

(3 marks)

3 A lens is used to form an image of an object that is 2.5 cm high. For each part, write down the equation you use, and then show clearly how you work out your answer.

(a) Calculate the magnification when the image is 10 cm high.

....

magnification = ... *(2 marks)*

(b) Calculate the magnification when the image is 2 cm high.

....

magnification = ... *(2 marks)*

4 A lens has a focal length of 3 cm. Draw a ray diagram to determine the magnification produced when the object is 5 cm from the lens.

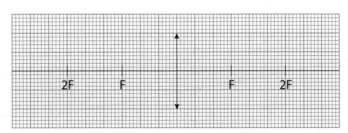

magnification = ... *(5 marks)*

Virtual images in lenses

1 The diagram shows the image formed when an object is viewed through a converging lens.

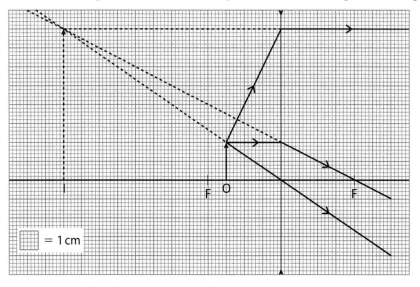

☐ = 1 cm

> **Guided**

(a) Explain why some rays are shown as dashed lines.

The dashed line shows where the rays ....

.... *(2 marks)*

(b) Describe the image shown in the diagram.

....

.... *(3 marks)*

(c) Calculate the magnification of the image. Write down the equation you use, and then show clearly how you work out your answer.

....

....

magnification = ... *(2 marks)*

2 An object is 40 cm from a concave lens with a focal length of 24 cm. Complete the ray diagram to find the distance of the image from the lens. The diagram is drawn to scale.
The scale is 1 cm : 4 cm.

> Remember that a concave (diverging) lens makes light from a point spread out (diverge) more.

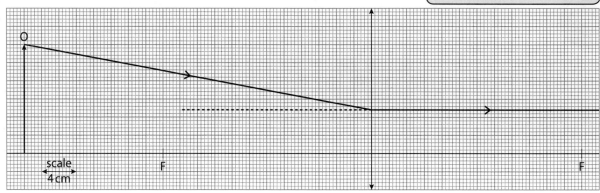

distance of image from lens = ... cm   *(4 marks)*

Make sure you draw ray diagrams as accurately as you can.

The eye

G-E **1** Label the diagram of the eye using the following terms.

| ciliary muscles | cornea | iris | lens | pupil | retina |

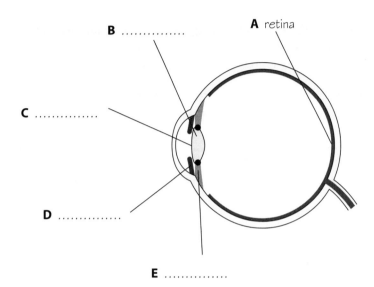

B ....

A retina

C ....

D ....

E ....

(3 marks)

> Check where the lines are pointing. Make sure that you can correctly label a diagram of the eye and that you know the function of each part of the eye.

D-C **2** Name the parts of the eye that:

(a) focus light onto the retina .... *(1 mark)*

(b) control the amount of light entering the eye .... *(1 mark)*

(c) change the shape of the lens. .... *(1 mark)*

B-A* **3** Some prescription drugs can affect the muscles that change the size of the iris. Suggest how a person's vision would be affected if the muscles in the iris could not move.

....

....

....

.... *(4 marks)*

> When an exam question says **suggest**, it means that you may not have been taught the answer. You should be able to work it out from what you have been taught.

Range of vision

G-E **1** Which of the following is the near-point distance for the average adult human eye?
Tick (✓) the box next to the correct answer.

☐ A: 25 km   ☐ B: 25 m   ☐ C: 25 cm   ☐ D: 25 mm

(1 mark)

D-C **2** The normal adult human eye has a far-point that is at infinity. Explain what this means.

Guided

> Infinity is bigger than the biggest number you can think of, but here you can consider it as just meaning very big.

There is no limit to the distance of an object ....

.... *(2 marks)*

D-C **3** **(a)** Describe **two** similarities between the eye and a camera.

....

....

.... *(2 marks)*

(b) Describe **two** differences between the eye and a camera.

....

....

.... *(2 marks)*

> When you are asked to describe similarities or differences, you must write about both of the things you are comparing.

B-A* **4** A person looking at a tablet computer looks up to see something in the distance. Describe what adjustments take place in the eye to produce a sharp image of the distant object.

....

.... *(2 marks)*

Correction of sight problems

1 The statements below are about long sight and short sight. Tick (✓) **one** box for each statement to show whether it applies to a person with short sight or long sight.

Statement	Short sight	Long sight
A: The images of objects close to the student form a sharp image on the retina.		
B: The student's eyeball is shorter than it should be.		
C: The student's lens cannot become curved enough to correct the problem.		
D: The student wears spectacles with diverging lenses to correct the problem.		

(4 marks)

> Make sure that you know the difference between short sight and long sight; what causes the two defects; and how the defects can be corrected.

2 (a) An optician gives a patient spectacles to improve their vision. The diagram shows the type of lens in the spectacles. Explain how the lenses help the patient.

> Guided

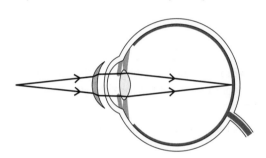

The .... lens refracts the light so that a sharp image of an

object that is .... *(2 marks)*

(b) Suggest **two** reasons why the patient may prefer contact lenses to spectacles.

....

.... *(2 marks)*

> Contact lenses rest on the cornea and cover the whole pupil.

3 A child needs to wear spectacles with a concave lens in order to see a sharp image of a distant object. Complete the diagram to show how the lens is able to correct the child's vision.

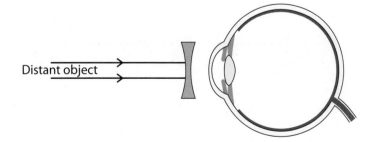

Distant object

(3 marks)

Power of a lens

 1 A student has two magnifying glasses. One is more powerful than the other. Tick (✓) **two** boxes to suggest the correct reasons why the magnifying glasses have different powers.

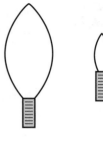

Reason	Tick (✓)
A: One lens is wider than the other.	
B: They are made of different transparent materials.	
C: They have different curvatures.	
D: The thicker lens is stronger.	

(2 marks)

 2 A lens has a focal length of −20 cm.

 (a) Calculate the power of the lens in dioptres. Write down the equation you use, and then show clearly how you work out your answer.

focal length: −20 cm is −0.20 m

$power = \dfrac{1}{f} =$ ....

power = ... dioptres   *(3 marks)*

To calculate the power in dioptres the focal length of the lens must be in metres. Make sure you understand what the negative sign means in front of the focal length.

(b) What type of lens has a negative focal length?   ....   *(1 mark)*

 3 An optician suggests that a patient needs spectacle lenses with a power of 2.5 dioptres.

HIGHER

(a) Calculate the focal length of the lenses. Write down the equation you use, and then show clearly how you work out your answer. Give the unit in your answer.

....

....

focal length = ... unit ....   *(2 marks)*

(b) The patient decides to pay for more expensive lenses that are made out of a high refractive index material. Explain the benefits this type of material will provide in spectacle lenses.

....

....

....

....   *(4 marks)*

Total internal reflection

  1 Complete each of the diagrams showing what happens to most of the light ray when it hits the boundary.

     (a) Diagram A: The angle of incidence is less than the critical angle.   *(2 marks)*

     (b) Diagram B: The angle of incidence is greater than the critical angle.   *(2 marks)*

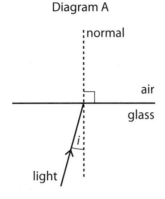

Diagram A   Diagram B

> Don't forget to label your lines. Remember that total internal reflection only occurs when light is going from glass or water into air.

  2 A diver swimming underwater can see the image of an object on the bottom of a swimming pool. The image appears to be above the surface of the water. Explain why this happens.

  Light from the object on the bottom hits the surface at an angle ....

.... and is ....

The light appears to come ....   *(3 marks)*

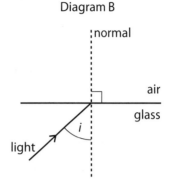 **3** **(a)** Polycarbonate is a transparent polymer used in safety glasses. The critical angle in polycarbonate is 39.12°.

       Calculate the refractive index of polycarbonate. Write down the equation you use, and then show clearly how you work out the answer.
Give your answer to an appropriate number of significant figures.

> You will need a scientific calculator for these questions.

     ....

     ....

     ....

                                      refractive index = ...   *(4 marks)*

     (b) The sparkle of a cut diamond is partly due to the very high refractive index of 2.4. Calculate the critical angle for diamond.

     ....

     ....

                                        critical angle = ...   *(2 marks)*

Had a go ☐ Nearly there ☐ Nailed it! ☐

Other uses of light

G-E **1** The diagram shows a length of optical fibre. Complete the diagram to show how the ray of light travels along the fibre.

Guided

The first arrow has been added for you as part of the guidance.

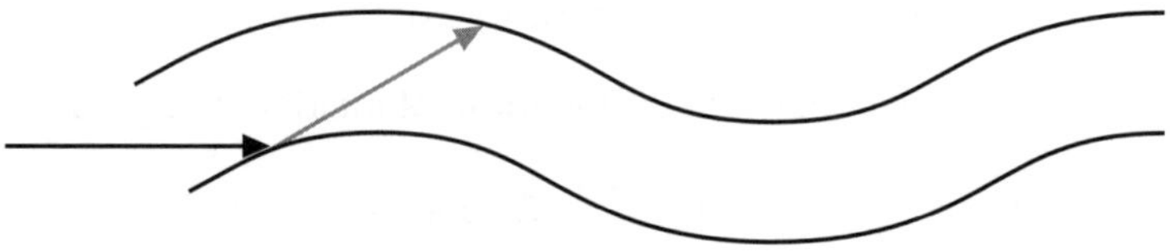

(2 marks)

D-C **2** **(a)** Explain why light travels along an optical fibre without escaping from the sides.

Remember that optical fibres are strands of solid glass.

....

.... *(2 marks)*

(b) Endoscopes are used by doctors to examine inside the digestive system of people. The endoscope has a long thin cable containing a bundle of optical fibres that can be pushed down the patient's gullet. The doctor looks through an eyepiece at one end of the cable.

Explain how the doctor is able to see an image of the inside of the patient's stomach.

Remember that the inside of the stomach is dark.

....

....

.... *(2 marks)*

B-A* **3** Lasers are used in eye surgery to correct long- and short-sightedness.

(a) State the properties that lasers have that enable them to be used in this way.

....

.... *(2 marks)*

(b) Explain how laser surgery corrects short-sightedness.

....

....

....

.... *(4 marks)*

Physics six mark question 1

Many people have either long sight or short sight. Explain the causes of these problems and how they can be treated.

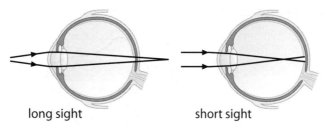

long sight   short sight

> You will get better marks on extended writing questions if you plan your answer before you start writing. In this question you need to think about how images are formed in the eye. Some questions to think about are:
> - What happens to light when it enters the eye?
> - What makes someone short or long sighted?
> - What methods are available to correct long and short sight?
> - How do these methods work?

....

....

....

....

....

....

....

....

....

....

....

....

....

....

....

....

....

....

....

....

.... *(6 marks)*

Centre of mass

1 A student wants to find the centre of mass of a sheet of plywood cut into an irregular shape. The instructions for this are given below but the sentences are mixed up. Put the sentences in the correct order and write down the order of the letters.

A: Move the thread to two other points on the object and repeat the method.

B: Hang a plumb line from the same hook.

C: Mark the line of the plumb line on the object.

D: Mark the point where the lines cross.

E: Fix a thread to the object and hang it from a hook.

> You should be able to follow the instructions if you have them in the correct order.

    Answer: *E*.........................

(2 marks)

2 State what is meant by the **centre of mass** of an object.

....

.... *(2 marks)*

3 In a DIY store, cut out digits for fixing to the front doors of houses are hung on hooks. Most of the digits hang lopsidedly when hung from a hole in the top of the digit.

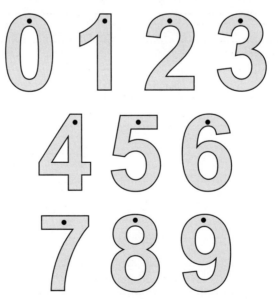

(a) Which digits will hang correctly?   .... *(1 mark)*

(b) Explain why most of the digits hang lopsidedly.

> Make sure you refer to centre of mass in your answer.

....

.... *(2 marks)*

The pendulum

G-E **1** Which of the following describes a complete swing of a pendulum? Tick (✓) the box next to the correct answer.

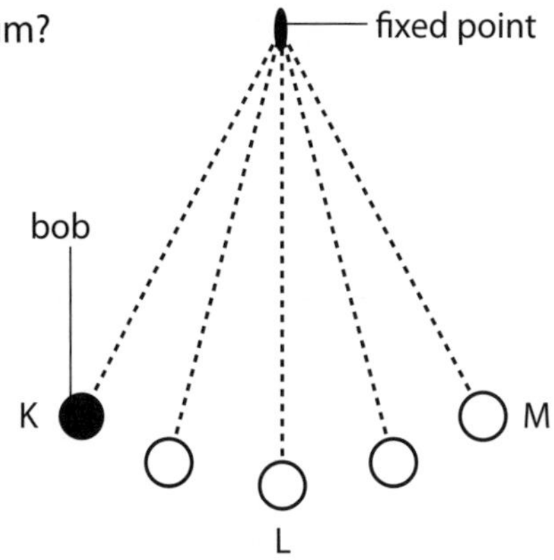

☐ A: from K to M

☐ B: from K to M and back

☐ C: from L to K and back

☐ D: from L to M and back

(1 mark)

D-C **2** A long rope has been tied over the branch of a tree to be used as a swing.

(a) A child on the swing makes 24 complete swings in 60 s.

(i) Explain why it is a good idea to measure the time for more than one swing.

....

.... *(2 marks)*

(ii) Calculate the frequency of the swing and give the unit.

> Frequency is the number of swings in 1 second. Remember that the period of a pendulum is the time for 1 complete swing.

....

....

frequency = ... unit .... *(3 marks)*

> **Guided**

(b) Calculate the period of the swing. Write down the equation you use, and then show clearly how you work out the answer.

T = 1/f = ...

....

period = ... s *(2 marks)*

(c) The children want to change the period of the swing. Suggest **one** change they could make.

> You do not have to say how the period will change.

....

(1 mark)

B-A* **3** The period of a pendulum is 0.75 s. Calculate the frequency of the pendulum. Write down the equation you use, and then show clearly how you work out the answer.

....

....

frequency = ... Hz *(2 marks)*

Turning effect and levers

G-E 1   The diagram shows a workman using a steel pole to lever a rock out of the ground.

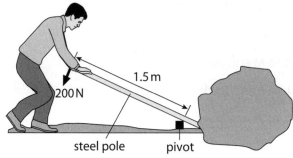

200 N   1.5 m

steel pole   pivot

(a) When the workman pushes with a force of 200 N at right angles to the pole, the rock starts to move. Show that the moment of the force is 300 Nm. Write down the equation you use, and then show clearly how you work out the answer.

....

.... *(2 marks)*

(b) Which of the following could the workman do to make it easier to move the rock?
Tick (✓) **two** boxes.

☐   A: use a longer pole   ☐   C: move the pivot away from the rock

☐   B: hold the pole closer to the rock   ☐   D: move the pivot closer to the rock

(2 marks)

D-C 2   The diagram shows two children on a seesaw.
Show that the seesaw is balanced.

Guided

Write down the equation you use, and then show clearly how you work out the answer.

2.4 m   1.8 m

150 N   200 N

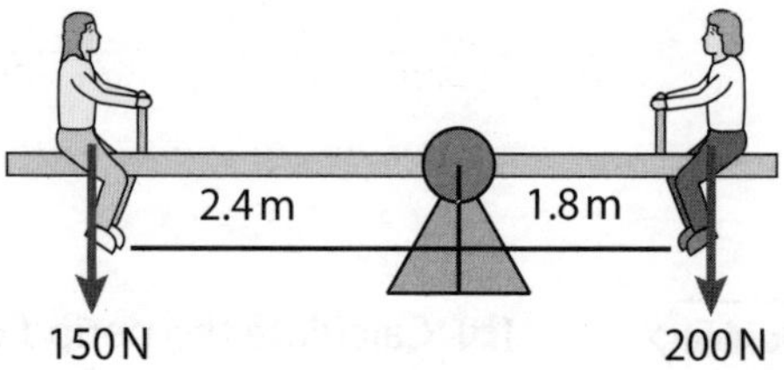

$M = F \times d$

clockwise moment = ...

anticlockwise moment = ...

The seesaw is balanced because .... *(3 marks)*

B-A* 3   Look at the diagram in question **1**.

(a) Explain why the steel pole being used as a lever is a **force multiplier**.

....

.... *(2 marks)*

(b) The rock is just about to move when a force of 200 N is used and the horizontal distance of the pivot to the rock is 0.2 m. Calculate the force on the rock when it starts to move. Write down the equation you use, and then show clearly how you work out the answer.

> If the rock is just about to move it is balanced on the steel pole.

....

....

....

force on rock = ... N   *(3 marks)*

Moments and balance

This whole page is Higher material.

For all questions on this page, write down the equation you use, and then show clearly how you work out the answer.

1 A child weighing 150 N is sitting at the end of a seesaw that is 3.2 m long and pivoted at its midpoint. An adult weighting 600 N sits on the seesaw. Calculate the distance from the pivot (d) of the adult when the seesaw is balanced.

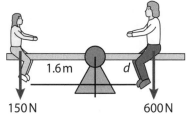

1.6 m   d

150 N   600 N

If the seesaw is balanced then clockwise moment

about the pivot = anticlockwise moment about the pivot

clockwise moment = M = F × d =

....

....

d = ... m   *(4 marks)*

2 A fireman uses a crowbar to force open a door to a burning building. The crowbar is 1.2 m long and pivoted 0.2 m from one end. Calculate the maximum force the fireman can apply to the door when a force of 400 N is applied at right angles to the end of the crow bar.

Remember that the moment is the force multiplied by the distance from the pivot perpendicular to the line of action of the force.

....

....

....

maximum force = ... N   *(3 marks)*

3 A uniform plank 2.8 m long with a weight of 180 N is leaning against a vertical fence. One end of the plank rests on the ground 2 m from the fence. The other end of the plank rests on the fence at a height of 2 m. The fence will collapse when the reaction force of the fence on the plank exceeds 100 N. Show that the fence will not collapse.

EXAM ALERT

Remember to show all the stages of your calculations.

Students have struggled with questions like this in recent exams – **be prepared**

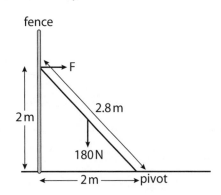

fence

F

2 m

2.8 m

180 N

2 m   pivot

Uniform means that the plank is symmetrical. Remember that weight always acts vertically. Draw forces or distances on the diagram if it helps you to work out how to answer the question.

....

....

....   *(3 marks)*

Stability

G-E   **1**   At breakfast a student puts three packets of cereal on a table as shown in the diagram.

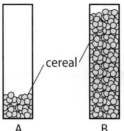

cereal

The fuller the packet the higher the centre of mass.

A   B   C

(a) Put the packets in order of stability with the most stable first.

.... *(1 mark)*

(b) Explain the order you have given.

....

.... *(2 marks)*

D-C   **2**   A school is providing new, safer stools for its laboratories that do not topple over easily.

AQA SKILL
Suggest
Page 98

(a) Suggest a design for the stools.

....

.... *(2 marks)*

⟩ **Guided** ⟩   **(b)** Explain why your design of stool is less likely to topple.

Only when the stool is tilted at a large angle does the line of action of the

.... *(2 marks)*

B-A*   **3**   A roly-poly toy like the one shown will always return to an upright position if it is pushed over.
HIGHER   X marks the position of the centre of mass. Explain the design of the toy. Sketch a diagram to illustrate your answer.

X

Use the idea of moments in your answer.

....

....

.... *(3 marks)*

Hydraulics

1 Use the correct words from the box to complete the sentences.

| compressed | dissipated | exerted | poured | reflected | transmitted |

A hydraulic system is filled with a liquid because liquids cannot be .... The pressure

at any point in the liquid is .... equally in all directions. A force .... on the

liquid at one point will be passed to other points in the liquid. *(3 marks)*

In hydraulics, remember that the same
pressure is transmitted through the liquid, not
the same force – the force may well be larger.

Students have struggled with questions
like this in recent exams – **be prepared!**

2 The diagram shows a simple hydraulic system that can be used to lift loads that are much bigger
than the effort needed. Explain how the hydraulic system works as a force multiplier.

> Guided

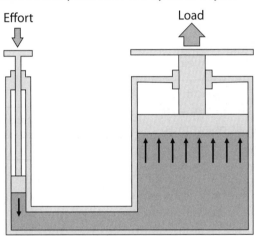

Effort   Load

Force multiplier means that
the load lifted is bigger than
the effort used.

The effort acting over the area of the small piston produces a .... This is

transmitted ....

The load is bigger than the effort because ....

.... *(3 marks)*

3 A hydraulic system has an effort piston with area $4.0 \times 10^{-4} \, m^2$ and
load piston with area $8.0 \times 10^{-2} \, m^2$. The hydraulic system is used
to raise a load of 3.2×10^4 N.

Take care using scientific
notation in calculations.

(a) Calculate the pressure in the hydraulic system. Write down the equation you use, and then
show clearly how you work out the answer. Give the unit for your answer.

....

....

pressure = ... unit .... *(3 marks)*

(b) Calculate the effort required. Write down the equation you use, and then show clearly how
you work out the answer.

....

....

effort = ... N *(2 marks)*

Circular motion

G-E **1** A force is needed to make an object move in a circle.

(a) State the general name for this force.

.... *(1 mark)*

(b) Give the name of the type of force that makes the following objects move in a circle. Use words from the box.

| friction | gravity | tension | weight |

(i) A communication satellite in orbit around the Earth. ....

(ii) A car going around a roundabout. ....

(iii) The ball in the hand of a bowler in cricket swinging their arm over. ....

(3 marks)

D-C **2** Explain why a continuous force is needed to make a weight on a rope move in a circle over your head even though the speed of the weight is constant.

EXAM ALERT

Remember **centripetal** is not a type of force like friction or tension – it is the name given to the resultant force that causes an object to go in a circular path.

Students have struggled with questions like this in recent exams – **be prepared!**

....

....

....

....

....

.... *(3 marks)*

B-A* **3** Explain the following observations.

Guided

(a) An athlete running in the inside lane of an athletics track finds it more difficult to stay in lane on the bends than a runner in the outside lane moving at the same speed.

The runner on the inside track needs more ....

because .... *(2 marks)*

(b) Warning signs on sharp bends of roads give the maximum speed with which vehicles can safely take the turn.

....

.... *(2 marks)*

(c) A fully loaded lorry is more likely to skid when travelling around a bend than an empty lorry travelling at the same speed.

....

.... *(2 marks)*

Physics six mark question 2

Cranes similar to the one shown in the diagram are used on building sites to move heavy loads. The load is lifted and moved horizontally before being lowered to where it is needed. Occasionally cranes topple over, but safety inspections are supposed to prevent this happening. Discuss the reasons why cranes may topple over and the precautions taken to prevent this happening.

> You will get better marks on these questions if you plan your answer before you start writing.
>
> In this question you need to think about the factors that affect the stability of objects. Some questions to think about are:
> - What forces act on the crane?
> - What causes the crane to topple over?
> - How can the crane be made more stable?

....

....

....

....

....

....

....

....

....

....

....

....

....

....

....

....

....

....

....

.... *(6 marks)*

Electromagnets

EXAM ALERT

G-E 1 Two students made an electromagnet by winding wire around an iron nail. They connected the electromagnet to a battery and counted the number of steel paper clips it would pick up. Their results are shown in the table.

Number of turns of wire	Number of paper clips picked up
5	4
10	8
15	12
20	16

(a) Which **one** of the following is the best conclusion the students could draw from their experiment? Tick (✓) **one** box.

☐   A: The strength of the electromagnet only depends on the number of turns of wire.

☐   B: The strength of the electromagnet increases as the number of turns of wire increases.

☐   C: The strength of the electromagnet doubles when the number of turns of wire is doubled.

☐   D: The strength of the electromagnet varies with the number of turns of wire.

(1 mark)

AQA SKILL Explain Page 98

(b) State another factor that could affect the strength of the electromagnet.

.... *(1 mark)*

> Remember the factors that affect the strength of an electromagnet.

> Students have struggled with questions like this in recent exams – **be prepared!**

D-C 2 The diagram shows a loudspeaker.

Guided

Explain why the paper cone vibrates when a varying current flows through the coil.

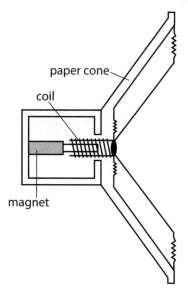

paper cone

coil

magnet

The paper cone attached to the coil vibrates as the force ....

.... because ....

.... *(2 marks)*

B-A* 3 A recycling works is investigating methods of sorting iron and steel objects by weight. Engineers suggest using an electromagnet or permanent magnet to make items of different weights fall into different channels. Evaluate the use of electromagnets for this purpose.

....

....

.... *(3 marks)*

The motor effect

1 The diagram shows an electric motor. Draw a ring around the correct answer to complete each sentence.

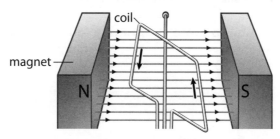

coil

magnet

N   S

> Make sure you can use Fleming's left-hand rule to decide the direction of the force on a current-carrying conductor in a magnetic field.

In the diagram the

| magnet |
| coil |
| pivot |

turns because the magnetic field and current are at

| 90° |
| 180° |
| 0° |

.

There is a force on the left-hand side of the coil of wire that makes it move

| sideways |
| down |
| up |

.

(3 marks)

2 Look at the diagram of the electric motor in question **1**.

(a) State **two** ways that the force on the coil can be increased.

....

....
(2 marks)

(b) State **two** ways that the direction of rotation of the coil can be reversed.

....

....
(2 marks)

> **Guided**

(c) Explain why there is no resultant force on the coil when it is in the vertical position.

The force on the top of the coil is upwards and the force on the bottom of the coil is

.... The forces are equal and opposite because

....
(2 marks)

3 Meters in which a needle moves are constructed as in the diagram.
Explain how the meter gives a steady reading when the current is steady.

> Think about what would happen if the restoring spring was not attached.

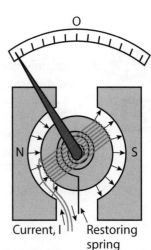

O

N   S

Current, I   Restoring spring

....

....

....

....

....

....
(2 marks)

Had a go ☐   Nearly there ☐   Nailed it! ☐

Electromagnetic induction

G-E   **1**   Label the diagram using words from the box.

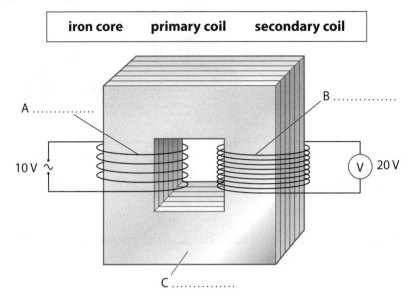

| iron core | primary coil | secondary coil |

A ....

B ....

10 V ⏦   Ⓥ 20 V

C ....

(2 marks)

D-C   **2**   The diagram shows a magnet and coil.

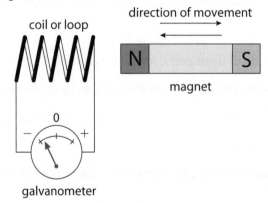

coil or loop

direction of movement

N   S

magnet

0

galvanometer

(a) Describe what you would see on the meter when the magnet is moved in and out of the coil.

....

.... *(2 marks)*

〉 **Guided** 〉   **(b)** Explain your observation of what happens when the magnet moves in one direction.

Try to use the term **induced** in your answer.

When the magnet moves the coil 'cuts' ....

.... *(2 marks)*

B-A*   **3**   Explain what happens when an alternating potential difference is connected to the primary coil of a transformer.

Remember that a conductor must be in a changing magnetic field for a potential difference to be induced across it.

Students have struggled with questions like this in recent exams – **be prepared!**

....

....

.... *(3 marks)*

Step-up and step-down transformers

1 Draw a ring around the correct answer to complete each sentence.

A step-up transformer has

| fewer |
| the same |
| more |

turns of wire on the secondary coil than the primary coil.

The potential difference across the secondary coil is

| greater than |
| the same as |
| less than |

across the primary coil.

If the potential difference across the secondary coil is less than across the primary coil it is

a

| step-up |
| step-down |
| relay |

transformer.

> Make sure you know what step-up and step-down transformers do.

(3 marks)

2 The diagram shows a transformer

Show that the potential difference of the output is 50 V.

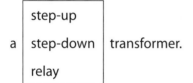

The number of turns on the secondary

coil is 5 times ....

....

so the output on the secondary coil

....

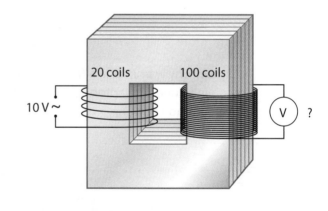

20 coils   100 coils

10 V ~

V   ?

> The transformer equation given on page 97 could help you with the answer.

(2 marks)

B-A* **3** In the National Grid, electricity is distributed around cities at 33 000 V. Local transformers convert this to 230 V for distribution to homes. The houses served by a transformer draw a total current of 800 A.

(a) Calculate the current carried by the cable to the transformer. Write down the equation you use, and then show clearly how you work out the answer.

....

....

Current = ... A   *(2 marks)*

(b) State the assumption you made in your calculation.

....   *(1 mark)*

Had a go ☐ Nearly there ☐ Nailed it! ☐

Switch mode transformers

G-E

Guided

1 The table below gives some statements about traditional transformers with primary and secondary coils, and modern switch mode transformers. Tick (✓) the boxes to show which type of transformer the statement applies to.

Some statements will apply to both types of transformer.

Statement	**Traditional transformer**	**Switch mode transformer**
works at mains frequency of 50 Hz	✓	
small and light		
converts mains electricity (230 V) to a lower potential difference		
works at a frequency up to 200 kHz		
contains a heavy iron core	✓	

(3 marks)

D-C

2 A warning about adaptors and battery chargers is shown below.

Warning!

Turn it off.

Do not leave mains adaptors and battery chargers plugged into mains electricity.

Explain why warnings like this used to be included with mobile phone and laptop computer chargers but are less common now.

Traditional transformers can get very hot when plugged in.

....

....

.... *(3 marks)*

B-A*

3 Switch mode transformers have replaced older traditional transformers for powering and charging batteries in consumer electronic gadgets such as tablet and laptop computers and mobile phones. Suggest why traditional transformers have been replaced by switch mode transformers for these uses.

....

....

.... *(3 marks)*

Physics six mark question 3

The diagrams show a circuit breaker, which is used to cut the power in a circuit if the current becomes too large. The diagrams do not show the spring that pushes the switch away from the contacts and the spring that holds the soft iron bolt in place. Explain how a large current causes the circuit breaker to cut the power.

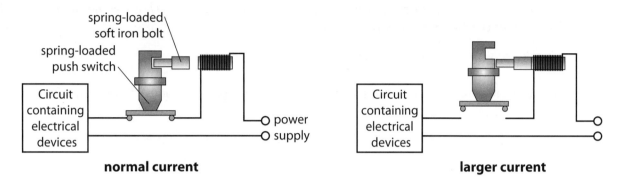

normal current   **larger current**

> You will get better marks on these questions if you plan your answer before you start writing.
>
> In this question you need to think about electromagnets. Some questions to think about are:
> - What is the purpose of the coil of wire?
> - What happens to the coil when the current increases?
> - What is the effect on the soft iron bolt?

....

....

....

....

....

....

....

....

....

....

....

....

....

....

....

....

....

....

.... *(6 marks)*

Further Additional Science Biology B3 practice paper

Time allowed: 60 minutes

This Practice Exam Paper has been written to help you practise what you have learned and may not be representative of a real exam paper.

This paper is a Higher Tier paper. Sample Foundation Tier papers can be found on the AQA website.

1   Substances can move into and out of cells by the processes of active transport, diffusion or osmosis.

Here are three definitions of processes for substances moving into and out of cells.

    A   movement of particles down a concentration gradient

    B   movement of water from a dilute solution to a concentrated one through a partially permeable membrane

    C   movement of particles against a concentration gradient

(a) Give the letter that describes the following processes:

    (i) diffusion = ... *(1 mark)*

    (ii) osmosis = ... *(1 mark)*

(b) Give **one** way in which active transport differs from diffusion.

.... *(1 mark)*

2   The diagram shows a cross-section through the human heart. The arrows show the route that blood takes through the heart.

(a) Name the blood vessels labelled **X** and **Y**.

    (i)   **X** = ... *(1 mark)*

    (ii)   **Y** = ... *(1 mark)*

(b) Other than the direction of blood flow, give **two** differences between the pulmonary artery and the pulmonary vein.

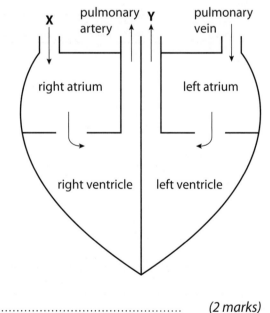

....

....

....

.... *(2 marks)*

(c) Explain why blood can only take the routes marked through the heart.

....

.... *(2 marks)*

3 A student investigates the loss of water from the leaves of a plant using the apparatus shown in the diagram.

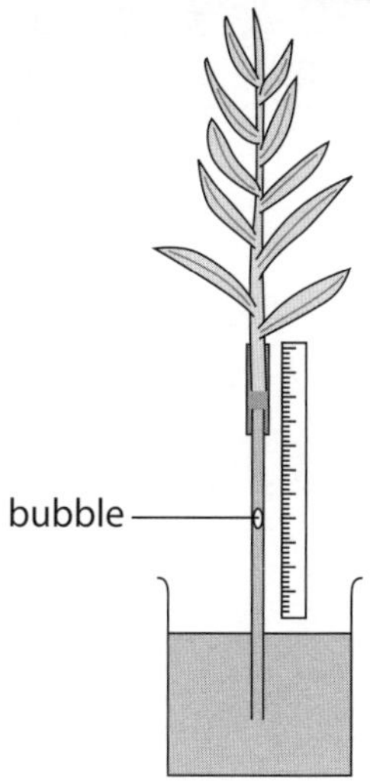

(a) What name is given to the loss of water from the leaves of a plant?

....

(1 mark)

(b) The student sets up the apparatus and measures how far the bubble moves in five minutes. He repeats the experiment to give four measurements.

bubble

Measurement	1	2	3	4
Distance moved by bubble in mm	59	68	21	62

Use appropriate results from the student's table to calculate the mean distance moved by the bubble. Give your answer in mm per minute.

....

....

mean distance = ... mm per minute   *(3 marks)*

(c) Explain what would happen to the distance moved by the bubble if the room became colder.

....

....   *(2 marks)*

(d) Explain why a decrease in light intensity in the room reduces the movement of the bubble.

....

....   *(2 marks)*

4 The diagram shows a villus, which is found in the digestive system.

surface epithelial layer

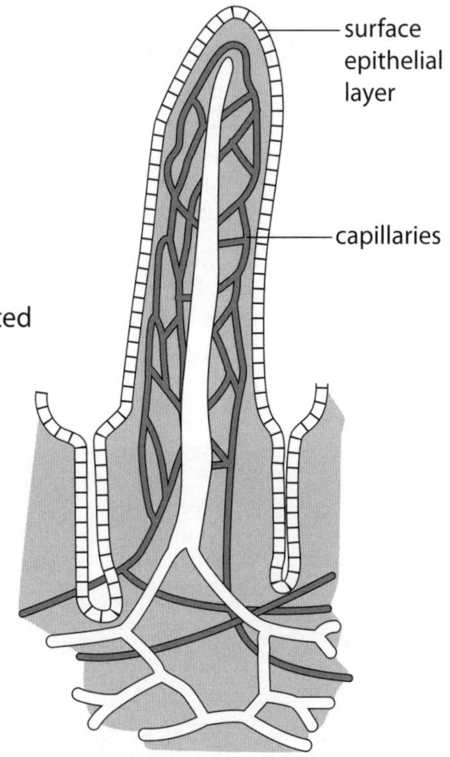

(a) Where are villi found in the digestive system?

....

....

(1 mark)

capillaries

(b) Explain **two** ways that the structure of the villus is adapted for the role that it performs in the digestive system.

....

....

....

....

....

....

....   *(4 marks)*

5 Scientists are worried that turtles may be affected by global warming. Turtles lay their eggs in sand on beaches. The temperature of the sand can affect the sex of the turtles that hatch from the eggs. The graph shows this relationship.

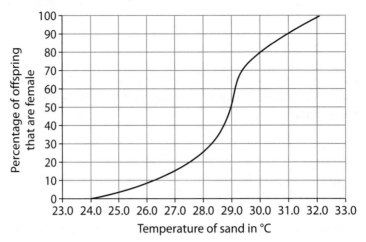

(a) (i) Use the graph to estimate the temperature of the sand that would give equal proportions of male and female turtles.

.... *(1 mark)*

(ii) A female turtle lays 120 eggs in sand at a temperature of 30 °C. Use the graph to estimate the number of male and female turtles that are likely to be produced from these eggs.

....

.... *(2 marks)*

(b) Use information from part **(a)** to suggest why scientists believe that global warming may affect the population of turtles.

....

.... *(2 marks)*

(c) Turtles live in the sea. Explain how the sea could help to reduce one of the causes of global warming.

....

....

.... *(3 marks)*

6 In healthy people, the concentration of glucose in the blood is maintained between 3.5 and 7.5 mmol per litre. The concentration of glucose in the blood rises after a meal, but returns to normal levels within 2 or 3 hours.

(a) The following passage describes the response of the body when a meal is eaten. Complete the passage by adding a suitable word in each space.

When a meal is eaten, the concentration of glucose in the blood rises. This rise in glucose

concentration results in the release of a hormone called ....

from cells in the .... *(2 marks)*

(b) Explain how the body responds if the concentration of glucose in the blood falls below 3.5 mmol per litre.

....

.... *(2 marks)*

(c) When the glucose concentration in the blood exceeds 9 mmol per litre, glucose starts to appear in the urine. This may be a sign that a person has diabetes. A person can be tested for diabetes. The person does not eat for eight hours and is then given a glucose drink. The blood glucose concentration is then monitored for two and a half hours.

(i) Suggest why the person does not eat for eight hours before this test.

.... *(1 mark)*

(ii) The results of this test for one person are shown in the graph.

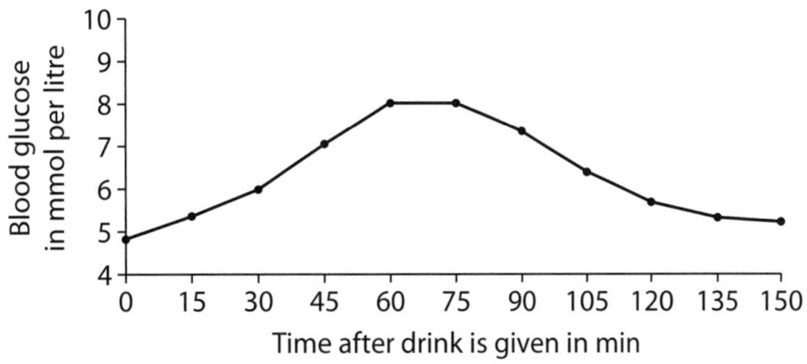

This person does not have diabetes. Explain how the results in the graph lead to this conclusion.

....

....

.... *(3 marks)*

(iii) What is the name of the hormone which releases glucose from stores in the body if the blood glucose levels fall too low?

.... *(1 mark)*

7 Scientists are trying to improve the substances used as artificial blood.

(a) The most important roles for a blood substitute are to transport oxygen and carbon dioxide around the body. Describe how these two substances are transported in the blood.

....

....

....

.... *(4 marks)*

(b) Artificial blood does not have the same ability as human blood to form clots. Describe how human blood clots.

....

.... *(2 marks)*

(c) Real human blood used for transfusions has antigens on the surface of blood cells. This means real blood must be carefully matched to the recipient's blood group. Use this information to suggest **one** advantage that artificial blood has over human blood for transfusions.

....

.... *(2 marks)*

8 During exercise, both the breathing rate and the body temperature rise. They then return to normal in the recovery period after exercise. These changes are shown on the graph.

(a) Describe the changes that happen in the thorax when we breathe in.

....

....

....

.... *(3 marks)*

(b) The total change in the temperature of the blood is less than 1°C. Explain how the body prevents the temperature from rising further.

....

....

....

.... *(4 marks)*

9 *In this question you will be assessed on using good English, organising information clearly and using specialist terms where appropriate.*

Animals are often used to produce food for humans. It is also possible to produce food for humans using microorganisms. Explain the advantages of using microorganisms for food production, rather than using animals.

....

....

....

....

....

....

....

....

....

....

....

....

....

....

....

....

....

....

....

....

....

....

....

....

....

....

....

....

....

.... *(6 marks)*

Further Additional Science Chemistry C3 practice paper

Time allowed: 60 minutes

This Practice Exam Paper has been written to help you practise what you have learned and may not be representative of a real exam paper.

This paper is a Higher Tier paper. Sample Foundation Tier papers can be found on the AQA website.

1 In the 1860s scientists were looking for order in the classification of the known elements. Two of the most important scientists working in this area were John Newlands in England and the Russian Dmitri Mendeleev.

(a) What was similar about the ways Mendeleev and Newlands organised the elements in their classification?

....

.... *(1 mark)*

(b) Explain **one** change that Mendeleev made to improve his classification.

....

.... *(2 marks)*

(c) In what order are the elements arranged in the modern periodic table?

.... *(1 mark)*

2 Group 1 elements in the modern periodic table are called the **alkali metals**.

(a) Suggest a reason why Group 1 elements are called the alkali metals.

.... *(1 mark)*

(b) The radii (a measure of atomic size) of Group 1 elements are shown in the table.

Element	Covalent radius in nm
Li	0.123
Na	0.157
K	0.203
Rb	0.216

Suggest the trend in atomic radii of Group 1 elements is linked to electron shells.

....

.... *(2 marks)*

(c) Descriptions of the reactions of the alkali metals with water are given in the table.

Element	Description of reaction with water
Li	fizzes about on surface of water
Na	melts into a ball and fizzes about
K	bursts into flames in a ball and fizzes about
Rb	explodes on contact with water

Explain the trend in the reactivity of the Group 1 elements.

....

.... *(2 marks)*

3 Soaps and detergents are chemicals used for cleaning. Soap is made from animal fat, and has been used for hundreds of years. Detergents are manufactured cleansing agents that were designed by chemists for use in hard water areas, where soaps didn't work very well. Most detergents are made from chemicals from the naphtha fraction from crude oil.

(a) Which ion is commonly found in hard water? Draw a ring around the correct answer.

Ca^{2+} Na^+ CO_3^{2-} *(1 mark)*

(b) Explain why soaps don't work well in hard water.

....

.... *(2 marks)*

(c) A student used the apparatus shown in the diagram to assess the relative hardness of three different samples of water.

The results of the investigation are shown in the table.

Water sample	1	2	3
Titration end-point in cm³	8.0	3.0	5.0

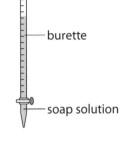

(i) Describe how the student would carry out the titration to compare the hardness of these water samples.

....

....

....

.... *(2 marks)*

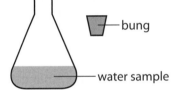

(ii) Each of the three samples of water were boiled for ten minutes and analysed again in the same way.

The results of the analysis on the boiled water are shown below.

Boiled water sample	1	2	3
Titration end-point in cm³	1.0	3.0	5.0

Explain which of the water samples can be described as temporary hard water.

....

.... *(2 marks)*

4 The equation for the combustion of ethene is shown below. It has been written to show the structural formulae of the reactants and products.

$$
\begin{array}{ccc}
\overset{\displaystyle H \quad H}{\underset{\displaystyle H \quad H}{C=C}} & + & \begin{array}{c} O=O \\ O=O \\ O=O \end{array}
\end{array}
\implies
\begin{array}{c} O=C=O \\ O=C=O \end{array}
+
\begin{array}{c} H \overset{O}{\diagup\diagdown} H \\ H \underset{O}{\diagup\diagdown} H \end{array}
$$

(a) Use the bond energies in the table to calculate the energy change for this reaction.

Bond	Bond energy in kJ
C—H	414
O=O	497
C=O	798
C=C	602
O—H	458

....

....

....

....

....

energy change = ... kJ   *(3 marks)*

(b) Explain in terms of bond energies why this reaction is exothermic.

....

....   *(2 marks)*

(c) Complete the balanced symbol equation for the combustion of butene (C_4H_8). State symbols
are **not** required.

C_4H_8 + ...O_2 → .... + ...   *(2 marks)*

5 Ethyl ethanoate is formed by the reaction of ethanol with ethanoic acid in the presence of a
catalyst of concentrated sulfuric acid.

ethanol + ethanoic acid ⇌ ethyl ethanoate

The energy level diagram for the formation of ethyl ethanoate is shown below.

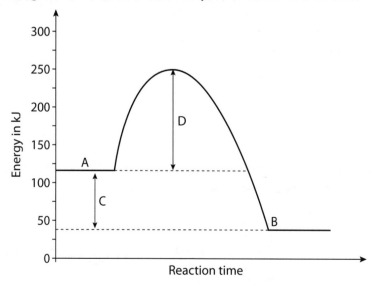

(a) Ethyl ethanoate is an ester. Suggest a use for ethyl ethanoate.

....   *(1 mark)*

(b) What does the energy change labelled 'D' on the graph represent?

....   *(1 mark)*

(c) Draw a dashed line on the graph to show the energy changes that would occur if the catalyst of concentrated sulfuric acid was **not** used.

(2 marks)

(d) By reference to the graph explain if the formation of ethyl ethanoate is exothermic or endothermic.

....

.... *(2 marks)*

6 As fossil fuels are finite and their use can cause pollution problems, we need to find new sources of energy. Two possible alternatives that are already being used are ethanol (C_2H_5OH) and hydrogen (H_2).

(a) Complete the following balanced equations for the combustion of hydrogen and ethanol.

(i) ....$H_2(g) + O_2(g) \rightarrow$ ....$H_2O(l)$ *(1 mark)*

(ii) $C_2H_5OH(g) +$ ....$O_2(g) \rightarrow$ ....$CO_2(g) + 3H_2O(l)$ *(1 mark)*

(b) Explain why burning hydrogen is potentially less harmful to the environment than burning ethanol.

....

.... *(2 marks)*

(c) Complete the missing information about ethanol in the table.

pH of solution	
Product of reaction with sodium metal	
Name of homologous series	

(3 marks)

(d) Ethanol can be converted into the compound shown below.

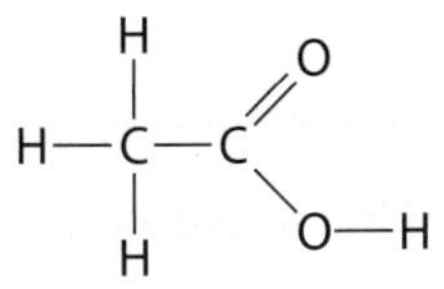

(i) State the name and the molecular formula for this compound.

.... *(2 marks)*

(ii) What kind of reaction changes ethanol into this compound?

.... *(1 mark)*

7 When a spark is applied to ammonia gas it decomposes into nitrogen and hydrogen.

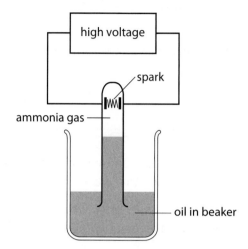

Under these conditions 95% of the ammonia decomposes according to the equation below.

$$2NH_3(g) \rightleftharpoons N_2(g) + 3H_2(g)$$

(a) What is the meaning of the $\rightleftharpoons$ sign, used in the above equation?

....

.... *(1 mark)*

(b) Explain what would happen to the amount of ammonia that decomposed if the gas pressure was increased.

....

.... *(2 marks)*

(c) Suggest a reason why oil is used in the beaker in this experiment rather than water.

....

.... *(1 mark)*

(d) Ammonia is manufactured in industry from nitrogen and hydrogen, by the opposite reaction:

$$N_2(g) + 3H_2(g) \rightleftharpoons 2NH_3(g)$$

(i) What is the name of this industrial process?

.... *(1 mark)*

(ii) The nitrogen for this process is obtained from the air. State the **two** substances used to make hydrogen for this process.

.... *(1 mark)*

(iii) In industry the reaction is carried out at a moderate temperature of 450 °C and a fairly high pressure of 200 atmospheres.

Explain why a lower temperature and higher pressure are not used even though they would produce a higher yield of ammonia.

....

....

.... *(2 marks)*

8 Some time ago the residents in one area of Cornwall reported that their water tasted funny. Analytical chemists working for the water authority carried out a series of tests. The results of some of the tests are shown below.

Test	Result
pH	6.0
flame test	slight yellow colour produced
addition of sodium hydroxide solution	white precipitate formed, which dissolves if excess sodium hydroxide is added
addition of hydrochloric acid	no visible change
addition of barium chloride/hydrochloric acid	white precipitate formed

The chemists concluded that the water contained excess aluminium sulfate, a chemical used in clearing small particles from the water during purifications.

(a) State which test showed the presence of aluminium ions and which test showed the presence of sulfate ions.

....

.... *(2 marks)*

(b) To find the amount of aluminium sulfate dissolved in the water the analysts had to know the concentration of the sodium hydroxide in mol/dm^3.

To do this they carried out a series of titrations to find the volume of 0.1 mol/dm^3 hydrochloric acid needed to neutralise 20 cm^3 portions of the sodium hydroxide. Their results are shown in the table below.

Titration	First level in cm^3	Second level in cm^3	Volume of HCl in cm^3
1st	0.0	12.2	12.2
2nd	12.2	22.2	10.0
3rd	22.2	32.0	9.8

Average volume of 0.1 mol/dm^3 HCl needed for neutralisation = 9.9 cm^3.

(i) Explain how the analyst would know when the correct volume of acid was added to neutralise exactly the 20 cm^3 portions of the sodium hydroxide.

....

.... *(2 marks)*

(ii) Why is the first titration result not used to work out the average volume of hydrochloric acid needed for neutralisation?

.... *(1 mark)*

(iii) Use the results to calculate the concentration of the sodium hydroxide solution in mol/dm^3.

....

....

....

concentration of sodium hydroxide solution = ... mol/dm^3 *(2 marks)*

9   *In this question you will be assessed on using good English, organising information clearly and using
specialist terms where appropriate.*

The river running through an industrial town has become polluted, and plants and aquatic life
have been damaged. It is thought that the pollution is caused by metal ions released by one of
the factories in the town. The analytical chemists working for the local water authority have been
asked to investigate the problem.

Explain a series of tests that you could do in a laboratory in an attempt to identify the polluting
metal ion. Include an assessment of the reliability of the tests.

....

....

....

....

....

....

....

....

....

....

....

....

....

....

....

....

....

....

....

....

....

....

....

....

....

....

.... *(6 marks)*

Further Additional Science Physics P3 practice paper

Time allowed: 60 minutes

This Practice Exam Paper has been written to help you practise what you have learned and may not be representative of a real exam paper.

This paper is a Higher Tier paper. Sample Foundation Tier papers can be found on the AQA website.

1   Some statements about X-rays and ultrasound used in medical diagnosis are shown in the table below. Put a tick (✓) in the boxes after each statement if the statement is true for X-rays, ultrasound or both.

Statement	X-rays	Ultrasound
They are a form of electromagnetic radiation		
They are a form of wave motion		
They can cause ionisation in cells		
They are reflected at the boundary between two different tissues		
They travel at the speed of light		

(5 marks)

2   Two students made some pendulums using lumps of modelling clay and some string. They used different masses of modelling clay and different lengths of string. Some of their results are shown below.

Test	Mass of modelling clay in g	Length of string in m	Time for ten swings in s
A	10	0.5	14.1
B	20	0.5	14.0
C	10	1.0	19.9
D	30	1.0	19.8
E	10	1.5	24.1
F	40	1.5	24.0

(a) Why did the students time ten complete swings for each test instead of just one?

.... *(1 mark)*

(b) Calculate the frequency of the pendulum in test F. Write down the equation you use, and then show clearly how you work out your answer.

....

....

frequency = ... Hz   *(3 marks)*

(c) (i) Which of the tests shown in the table above could the students use to show how the period of the pendulum depends on its length?

.... *(1 mark)*

(ii) What conclusion could the students make about the effect of the length of the pendulum on the period?

.... *(1 mark)*

(d) What do the results show about the effect of changing the mass of modelling clay on the period?

.... *(1 mark)*

3 The diagram shows an endoscope. It is made up of a bundle of optical fibres.

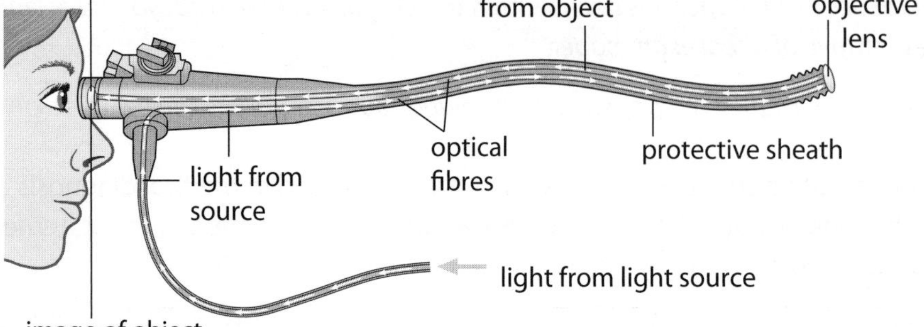

eyepiece lens

reflected light
from object

objective
lens

light from
source

optical
fibres

protective sheath

light from light source

image of object

(a) State **one** use for an endoscope.

.... *(1 mark)*

(b) When the endoscope is bent, light rays hit the sides of the optical fibres. Explain why the light does not escape from the optical fibres. You may draw a diagram to illustrate your answer.

....

.... *(2 marks)*

(c) The critical angle of the glass in the optical fibres is 40°. Calculate the refractive index of the glass. Write down the equation you use, and then show clearly how you work out your answer.

....

....

refractive index = ... *(2 marks)*

4 A converging lens can be used to produce a magnified image.

(a) Explain how light rays are refracted as they enter and leave the lens.

....

.... *(2 marks)*

(b) The focal length of the lens is 0.15 m. Calculate the power of the lens. Write down the equation you use, and then show clearly how you work out your answer.

....

....

power = ... dioptres *(2 marks)*

(c) In the diagram below the object is 0.1 m from the lens. Complete the ray diagram to determine the position and size of the image.

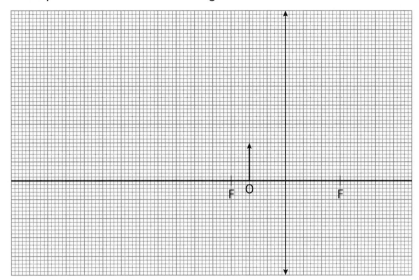

(3 marks)

(d) Calculate the magnification produced by the lens. Write down the equation you use, and then show clearly how you work out your answer.

....

....

Magnification = ... *(2 marks)*

5 The diagram shows a syringe containing a liquid with a cap over the needle.

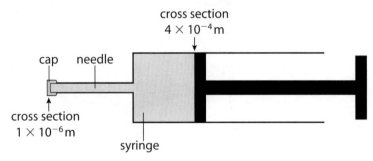

(a) Explain why liquid is forced against the cap on the needle when a force is applied to the end of the piston.

....

.... *(2 marks)*

(b) The piston is pressed with a force of 0.5 N. Calculate the force of the liquid on the cap on the needle. Write down the equation you use, and then show clearly how you work out your answer.

....

....

....

force = ...N *(3 marks)*

6 A potter designed the flower vase shown on the right.
On the diagram the centre of the X marks the centre of mass of the vase.

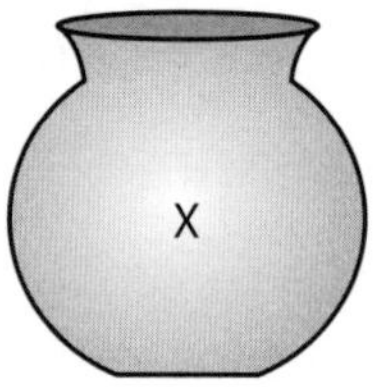

(a) What is meant by the term **centre of mass**?

.... *(1 mark)*

(b) The potter receives complaints that the vase topples over particularly when it has flowers in it.
Describe **two** changes the potter could make to the vase so that it is more stable.

....

.... *(2 marks)*

(c) Explain why the vase is more likely to topple over when it is nudged if it has a bunch of flowers in it.

....

....

.... *(3 marks)*

7 Electric vehicles have electric motors in each wheel that provide a force that turns the wheel.

(a) The diagram shows a simplified electric motor.

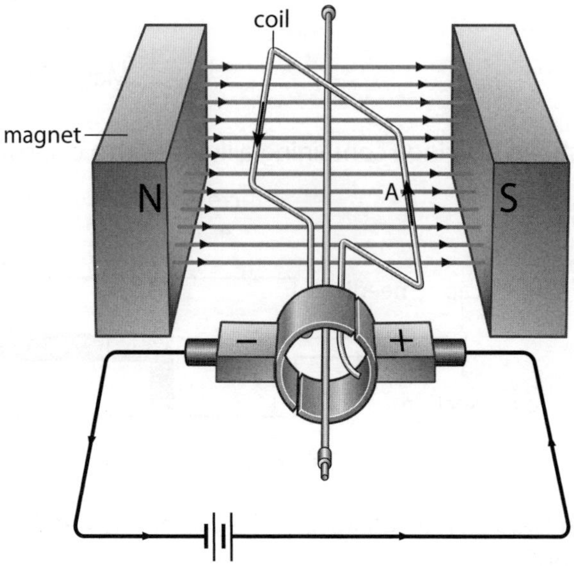

(i) Draw an arrow to show the direction of the force acting at point A on the coil. *(1 mark)*

(ii) There is a point in the rotation of the coil where there is no resultant force acting.
State when this occurs.

.... *(1 mark)*

(b) (i) State **two** ways that the force provided by the motors can be increased.

....

.... *(2 marks)*

(ii) What can be done to make the electric motors reverse the direction of the force on the wheels?

.... *(1 mark)*

8 A student uses a charger containing a transformer to recharge some batteries from the 230 V mains supply. The transformer has 1120 turns on the primary coil and 24 turns on the secondary coil. The current that recharges the batteries is 0.1 A.

(a) (i) How do you know from the description that the transformer in the charger is a step-down transformer?

.... *(1 mark)*

(ii) Calculate the potential difference across the rechargeable batteries. Write down the equation you use, and then show clearly how you work out your answer.

....

....

potential difference = ...V *(2 marks)*

(b) Calculate the current drawn by the charger from the mains supply. Assume that the transformer is 100% efficient. Write down the equation you use, and then show clearly how you work out your answer.

....

....

current = ...A *(2 marks)*

(c) The student decides to replace the charger with a new one. The new charger contains a switch mode transformer. Give **three** reasons why the student is likely to think the new charger is better.

....

....

.... *(3 marks)*

9 The diagram shows an asymmetrical mobile with two birds that are balanced by a weight. Calculate the weight of the ball that is needed to balance the mobile horizontally. Write down the equation you use, and then show clearly how you work out your answer.

....

....

....

....

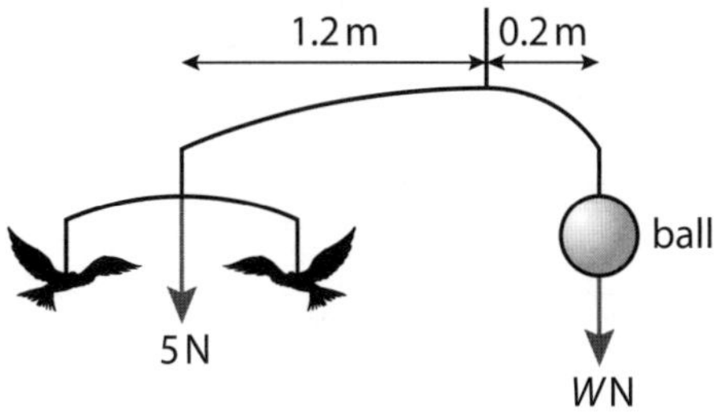

weight = ...N *(4 marks)*

10 *In this question you will be assessed on using good English, organising information clearly and using specialist terms where appropriate.*

The eye is often said to resemble a camera. Compare the function of the parts of the eye and the camera, and the formation of an image in both.

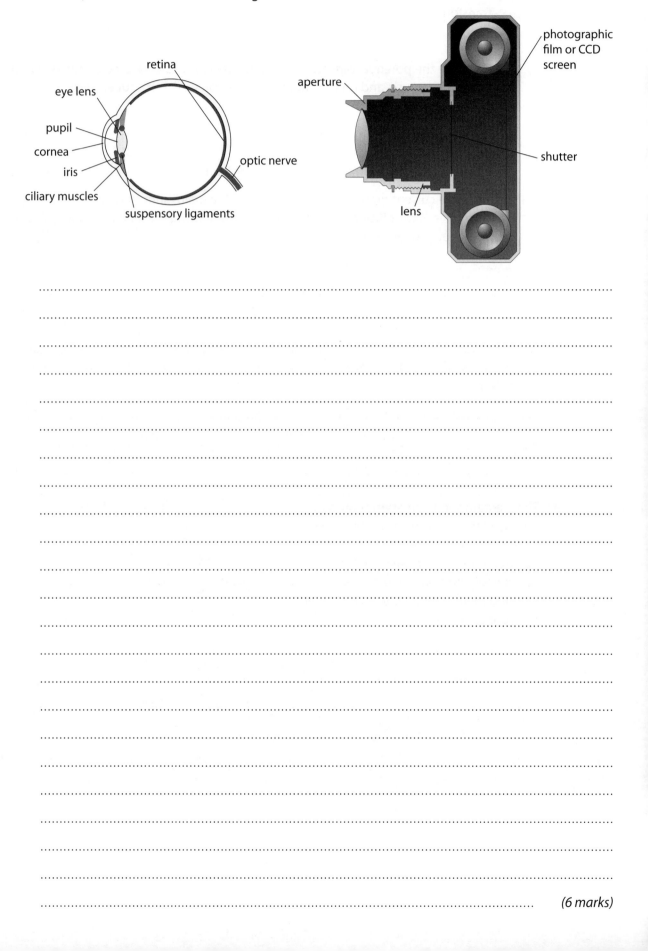

....

....

....

....

....

....

....

....

....

....

....

....

....

....

....

....

....

....

....

....

....

.... *(6 marks)*

Key

relative atomic mass
atomic symbol
name
atomic (proton) number

1	2											3	4	5	6	7	0
1 **H** hydrogen 1																	4 **He** helium 2
7 **Li** lithium 3	9 **Be** beryllium 4											11 **B** boron 5	12 **C** carbon 6	14 **N** nitrogen 7	16 **O** oxygen 8	19 **F** fluorine 9	20 **Ne** neon 10
23 **Na** sodium 11	24 **Mg** magnesium 12											27 **Al** aluminium 13	28 **Si** silicon 14	31 **P** phosphorus 15	32 **S** sulfur 16	35.5 **Cl** chlorine 17	40 **Ar** argon 18
39 **K** potassium 19	40 **Ca** calcium 20	45 **Sc** scandium 21	48 **Ti** titanium 22	51 **V** vanadium 23	52 **Cr** chromium 24	55 **Mn** manganese 25	56 **Fe** iron 26	59 **Co** cobalt 27	59 **Ni** nickel 28	63.5 **Cu** copper 29	65 **Zn** zinc 30	70 **Ga** gallium 31	73 **Ge** germanium 32	75 **As** arsenic 33	79 **Se** selenium 34	80 **Br** bromine 35	84 **Kr** krypton 36
85 **Rb** rubidium 37	88 **Sr** strontium 38	89 **Y** yttrium 39	91 **Zr** zirconium 40	93 **Nb** niobium 41	96 **Mo** molybdenum 42	99 **Tc** technetium 43	101 **Ru** ruthenium 44	103 **Rh** rhodium 45	106 **Pd** palladium 46	108 **Ag** silver 47	112 **Cd** cadmium 48	115 **In** indium 49	119 **Sn** tin 50	122 **Sb** antimony 51	128 **Te** tellurium 52	127 **I** iodine 53	131 **Xe** xenon 54
133 **Cs** caesium 55	137 **Ba** barium 56	139 **La** lanthanum 57	178 **Hf** hafnium 72	181 **Ta** tantalum 73	184 **W** tungsten 74	186 **Re** rhenium 75	190 **Os** osmium 76	192 **Ir** iridium 77	195 **Pt** platinum 78	197 **Au** gold 79	201 **Hg** mercury 80	204 **Tl** thallium 81	207 **Pb** lead 82	209 **Bi** bismuth 83	210 **Po** polonium 84	211 **At** astatine 85	222 **Rn** radon 86
223 **Fr** francium 87	226 **Ra** radium 88	227 **Ac** actinium 89	261 **Rf** rutherfordium 104	262 **Db** dubnium 105	266 **Sg** seaborgium 106	264 **Bh** bohrium 107	277 **Hs** hassium 108	268 **Mt** meitnerium 109	271 **Ds** darmstadtium 110	272 **Rg** roentgenium 111							

The lanthanides (atomic numbers 58–71) and the actinides (atomic numbers 90–103) have been omitted.

Elements with atomic numbers 112–118 have been reported but not fully authenticated.

Relative atomic masses for Cu and Cl have not been rounded to the nearest whole number.

Chemistry data sheet

Reactivity series of metals

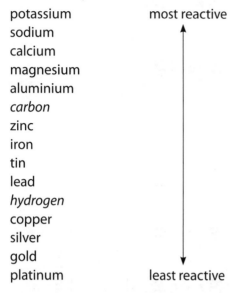

potassium   most reactive
sodium
calcium
magnesium
aluminium
carbon
zinc
iron
tin
lead
hydrogen
copper
silver
gold
platinum   least reactive

Elements in italics, though non-metals, have been included for comparison.

Formulae of some common ions

Positive ions		**Negative ions**	
Name	**Formula**	**Name**	**Formula**
hydrogen	H^+	chloride	Cl^-
sodium	Na^+	bromide	Br^-
silver	Ag^+	fluoride	F^-
potassium	K^+	iodide	I^-
lithium	Li^+	hydroxide	OH^-
ammonium	NH_4^+	nitrate	NO_3^-
barium	Ba^{2+}	oxide	O^{2-}
calcium	Ca^{2+}	sulfide	S^{2-}
copper(II)	Cu^{2+}	sulfate	SO_4^{2-}
magnesium	Mg^{2+}	carbonate	CO_3^{2-}
zinc	Zn^{2+}		
lead	Pb^{2+}		
iron(II)	Fe^{2+}		
iron(III)	Fe^{3+}		
aluminium	Al^{3+}		

Physics equations sheet

$s = v \times t$	s distance v speed t time
refractive index $= \dfrac{\sin i}{\sin r}$	i angle of incidence r angle of refraction
magnification $= \dfrac{\text{image height}}{\text{object height}}$	
$P = \dfrac{1}{f}$	P power f focal length
refractive index $= \dfrac{1}{\sin c}$	c critical angle
$T = \dfrac{1}{f}$	T periodic time f frequency
$M = F \times d$	M moment of the force F force d perpendicular distance from the line of action of the force to the pivot
$P = \dfrac{F}{A}$	P pressure F force A cross-sectional area
$\dfrac{V_p}{V_s} = \dfrac{n_p}{n_s}$	V_p potential difference across the primary coil V_s potential difference across the secondary coil n_p number of turns on the primary coil n_s number of turns on the secondary coil
$V_p \times I_p = V_s \times I_s$	V_p potential difference across the primary coil I_p current in the primary coil V_s potential difference across the secondary coil I_s current in the secondary coil

AQA specification skills

In your AQA exam there are certain **skills** that you sometimes need to **apply** when answering a question. Questions often contain a particular **command word** that lets you know this. On this page we explain how to spot a command word and how to apply the required skill.

Note: Watch out for our Skills sticker – this points out the questions that are particularly focused on applying skills.

Command word	Skill you are being asked to apply
Apply	You might be asked to **apply** what you know about a topic to a practical situation. For example: 'Apply your knowledge of series and parallel circuits to suggest the best wiring for a set of Christmas tree lights.'
Compare	**Compare** how two things are similar or different. Make sure you include both of the things you are being asked to compare. For example: 'A is bigger than B, but B is lighter than A.'
Consider	You will be given some information and you will be asked to **consider** all the factors that might influence a decision. For example: 'When buying a new fridge the family would need to consider the following things …'
Describe	**Describe** a process or why something happens in an accurate way. For example: 'When coal is burned the heat energy is used to turn water into steam. The steam is then used to turn a turbine …'
Discuss	In some questions you might be asked to make an informed judgement about a topic. This might be something like stem cell research. You should **discuss** the topic and give your **opinion** but make sure that you back it up with information from the question or your scientific knowledge.
Draw	Some questions ask you to **draw** or sketch something. It might be the electrons in an atom, a graph or a ray diagram. Make sure you take a pencil, rubber and ruler into your exam.
Evaluate	This is the most important one! Most of the skill statements start with **evaluate**. You will be given information and will be expected to use that information plus anything you know from studying the material in the specification to look at evidence and come to a **conclusion**. For example, if you were asked to evaluate which of two slimming programmes was better, then you might comment like this: 'In programme A people lost weight quickly to start with but then put the weight back on by the end of the sixth month. In programme B they did not lose weight so quickly to start with, but the weight loss was slow and steady and no weight was gained back by the end of the year. I therefore think that programme B is most effective.'
Explain	State what is happening and **explain** why it is happening. If a question asks you to explain then it is a good idea to try to use the word 'because' in your answer. For example: 'pH 2 is the optimal pH for enzymes in the stomach because the stomach is very acidic.'
Formulae	In some chemistry questions you will be expected to write chemical **formulae** for compounds. You will be given the symbols and the ions can be found on the Data Sheet on page 96.
Interpret	**Interpret** the data given to you on graphs, diagrams or in tables to help answer the question. For example: 'Use the data to show what happens when …'
Predict	A question may ask you to **predict** the outcome of a genetic cross or what will be formed at different electrodes in an electrolysis experiment. If you learn the patterns of genetic crosses and the rules about electrolysis then you will be able to do this.
Suggest	You will be given some information about an unfamiliar situation and asked to **suggest** an answer to a question. You will not have learned the answer – you need to **apply** your knowledge to that new situation. For example: 'I think that blue is better than green because …' or 'It might be because …'

Answers

You will find some advice next to some of the answers. This is written in italics. It is not part of the mark scheme but just gives you a little more information.

Biology answers

1. Into and out of cells

1 **(a)** dissolved (1)
  (b) down a concentration gradient (1)
2 **(a)** Osmosis is the movement of water molecules (1); across a partially permeable membrane (1); from areas of high water concentration to areas of low water concentration/from a dilute solution to a more concentrated solution (1).
  (b) (i) measure the mass of the Visking tubing and its contents before and after the experiment/ measure the change in volume in the beaker (1)
    (ii) They are too large (1); to pass through the small holes in the Visking tubing/partially permeable membrane (1).
3 **(a)** active transport (1)
  (b) to absorb mineral ions from the soil against the concentration gradient (1); plants use this process as they have to take up ions from the soil in very low concentrations/lower concentrations than in the plant (1)
  (c) Oxygen is needed for cells to respire (1); to provide the energy to make active transport happen (1).

2. Sports drinks

1 The drink is mostly composed of water (1); but will also contain dissolved ions/salts (1).
2 Balance of water, sugars and ions is important to keep the body functioning correctly (1); a drink with the correct balance will replace ions/sugars that are lost during sweating/respiration (1); rather than just replacing water (1).
3 **(a)** Drinking water will replace only the water lost/will not replace sugars or ions (1); so the fluid in the body's tissues will become more dilute/the body will work less efficiently (1).
  (b) (i) In general, using the drink increases the time that a person can exercise (1); although there are some people for whom the drink makes no difference/the drink decreases the time that a person can exercise (1).
    (ii) Most people who used the drink could exercise for longer (1); although one person exercised for a shorter period/three people found it had little or no effect on the period of exercise (1); and the sample size is too small to draw a valid conclusion (1).
    (iii) The study only looked at time of exercise and not recovery (1); so there is no evidence to support this claim (1).

3. Exchanging materials

1 **(a)** provide a large surface area (1)
  (b) capillaries (1)
2 **(a)** The surface of the small intestine is covered with villi (1). These help by increasing the surface area available for absorption (1).

(b) This makes the absorption of food molecules more efficient/effective (1); by reducing the distance that the molecules have to diffuse (1).
3 **(a)** An efficient blood flow helps to absorb the maximum amount of oxygen from the air (1); the flow of blood helps provide the body with the constant supply of oxygen it needs (1); and maintains the concentration gradient for diffusion (1).
  (b) Simple organisms often have a high surface area-to-volume ratio (1); and so get all the oxygen they need from simple diffusion (1); and are small enough that diffusion spreads the oxygen throughout the organism/whereas complex organisms need a mechanism for exchanging materials (1).

4. Ventilation

1 lungs (1); thorax (1); oxygen (1); carbon dioxide (1)
2 **(a)** diaphragm (1)
  (b) (i) ribcage/a rib (1)
    (ii) It helps to protect organs in the thorax of the body (1); especially the heart/lungs (1).
3 **(a)** Oxygen diffuses from the air into the blood through the alveoli (1); and carbon dioxide diffuses the other way/from the blood to the air in the alveoli (1).
  (b) Any four from: diaphragm moves down (1); ribs drawn upwards and outwards (1); this increases the volume inside the thorax (1); so pressure inside thorax reduces (1); air is drawn into lungs (1); this is breathing in/ inspiration (1). OR Any four from: diaphragm moves up (1); ribs move downwards and inwards (1); this decreases the volume inside the thorax (1); so pressure inside thorax increases (1); air is forced out of lungs (1); this is breathing out/exhalation (1).
  (c) They can be used to keep people alive when they cannot breathe for themselves (1), for example when they are unconscious/in a coma/under anaesthetic (1).

5. Exchange in plants

1 **(a)** water (1)
  (b) Carbon dioxide enters the leaves of the plant (1); by the process of diffusion (1).
2 Cells in the roots (1); are covered with root hairs (1); to increase the surface area that can be used for absorption (1).
3 **(a) (i)** guard cells (1)
    (ii) They control the size of the stomata in the leaf (1); and so alter the amount of gases that can enter and leave the leaf (1).
  (b) Gas exchange can occur more rapidly/efficiently (1), as the cells around air spaces have a larger surface area exposed to gases (1).
4 **(a)** The rate of evaporation was higher when the fan was on (1); because the movement of air removes water more quickly from the leaves (1).
  (b) The rate of evaporation becomes quicker than the rate at which it can take up water (1); the stomata of the plant close (1); to prevent evaporation from occurring (1).

6. The circulatory system

1 **(a)** circulatory system (1)
  (b) muscle (1)
2 **(a)** The two chambers are the right atrium (1); and the right ventricle (1).
  (b) valve (1); prevents the blood flowing the wrong way (1)

3 (a) aorta (1); vena cava (1)

(b) Usually arteries carry oxygenated blood and veins deoxygenated blood (1); but the pulmonary artery carries deoxygenated blood and the pulmonary vein oxygenated blood (1).

(c) Blood on the left-hand side of the heart is being pumped to all parts of the body (1); whereas blood from the right side of the heart only goes to the lungs (1); so the heart has thicker walls on the left side to pump the blood a greater distance (1).

7. Blood vessels

1 (a) The aorta carries oxygenated blood away from the heart (1).

(b) An artery has thick walls (1). These walls are composed of two types of fibres: muscular tissue (1) and elastic fibres (1).

2 The artery needs to be widened again (1); a stent can be used to do this (1).

3 (a) arteries (1)

(b) Capillaries are thin-walled/arteries are thick-walled (1); capillaries are much narrower than arteries (1); these features enable materials to pass between capillaries and tissues most easily (1); because most cells in the body are very close to a capillary (1).

4 (a) Veins have valves (1); to keep blood flowing in the same direction/prevent back flow (1).

(b) Veins have a thinner muscle wall than arteries (1); so it is easier to get the needle in (1). OR Veins contain blood under lower pressure (1); so taking blood is more controlled (1).

8. Blood

1 water (1); lungs (1); liver (1); kidneys (1)

2 (a) nucleus (1)

(b) (i) This pigment is called haemoglobin (1). Its role is to combine with oxygen in the lungs and carry it around the body (1).

(ii) Oxygen diffuses out of red blood cells into the tissues (1); because the oxygen concentration is lower in respiring tissues (1).

3 Platelets are cell fragments (1); they help the blood to clot in a wound/injury (1); and this prevents further loss of blood/prevents microorganisms getting in through the wound (1).

4 The man has developed a disease/an infection (1); white blood cells are produced in response to this infection (1); and help to destroy the microorganisms (1).

9. Transport in plants

1 (a) 3.2 cm (1) *Make sure you compare the same part of the bubble with the scale each time. In A, the bottom of the bubble is against the zero mark.*

(b) transpiration stream (1)

2 (a) The xylem is used to transport water (1) and mineral ions (1) in the plant from the roots to the leaves (1).

(b) (i) phloem (1)

(ii) The sugars are either taken to areas where the plant is growing (1); or taken for storage/to storage organs (1).

3 Water evaporates from the leaves (1); this helps to draw water up from the roots/through the xylem vessels (1); this is very important because the plant needs to replace water lost through transpiration (1); in order to photosynthesise/stop the plant from wilting (1).

10. Biology six mark question 1

A basic answer would only discuss one or two of the components of the blood. The answer would probably be limited to a description of the functions of the components.

A good answer would give three or four of the components of the blood. The answer would still contain mostly descriptions, but there may be some explanation for one or two of the blood components.

An excellent answer would give all the components of the blood. The answer would give descriptions, but there would be a good explanation of each component's role in keeping us healthy.

Examples of points made in the response:
- Red blood cell: contains haemoglobin; carries oxygen around the body; to supply to muscles/tissues; ensures that we can respire.
- White blood cell: part of body's defence mechanism; destroys microorganisms; to prevent infection or to return body to health after infection.
- Platelet: cell fragment that assists in clotting process; clots stop further loss of blood if body is damaged; and the clot acts as a barrier to prevent entry of microorganisms; preventing infection.
- Plasma: fluid in which blood cells are suspended; flows in blood vessels; transports soluble materials; such as carbon dioxide/products of digestion/urea.

11. Removing waste products

1 (a) carbon dioxide (1)

(b) lungs (1)

2 (a) in the blood (1)

(b) Urea is produced in the liver (1); by the breaking down of amino acids (1); urine is produced in the kidney (1) and is stored in the bladder before being excreted from the body (1).

3 (a) The kidney reabsorbs all the materials it needs from the blood (1); and releases the remains, along with urea and excess water, to make urine (1).

(b) A healthy kidney should remove ions from the blood (1); and should remove urea to be excreted (1); but a healthy kidney should also absorb glucose from the blood, so this kidney is not behaving properly (1).

(c) Any three from: blood/body fluids become more concentrated (1); this could mean that cells lose water into the blood/body fluids (1); by osmosis (1); and become damaged (1).

12. Kidney treatments

1 (a) between 14% and 16% (1)

(b) receiving a kidney transplant from a living relative (1)

2 Using a machine in hospital leads to difficulties keeping a job because: appointments need to be made (1); these will tend to be during the working day (1); and will happen frequently (1). OR Using a machine at home makes it easier to keep a job because: you can use the machine in your own time (1); so you can avoid using the working day (1); and have the flexibility to use it when you like, e.g. overnight (1).

3 (a) Blood flows through the machine inside partially permeable membranes (1); dialysis fluid contains glucose and amino acids in the same concentration as the blood, so these do not diffuse out of the patient's blood (1); but waste products like urea diffuse into dialysis fluid (1).

(b) If water were used as dialysis fluid, then the patient would lose sugars/amino acids/ions from their blood (1) via the process of diffusion (1) and take in water from the dialysis solution by osmosis (1).

4 Three marks for aspects – but no more than two from either section – and one mark for conclusion. Positive aspects: it is a long-lasting treatment (1); more convenient for patient than constant dialysis (1); drugs are available to prevent rejection (1). Negative aspects: long waiting lists (1); difficult to find the right 'match' of donor kidney to patient (1); kidney can be rejected by the host (1). Conclusion: for example, although there are potential disadvantages, having a long-lasting solution for kidney failure means that transplants are a positive treatment (1).

13. Body temperature

1 brain (1); sweat (1); water (1); food (1)
2 **(a)** temperature receptors in the skin (1)
  (b) Shivering is the process where muscles in the body (1) repeatedly contract (1); this process helps, because it releases energy (1).
  (c) Blood vessels supplying surface capillaries can constrict (1); which causes less blood to flow to the skin surface (1); so that less heat is lost (1).
3 Any four from: more sweat is produced (1); by the sweat glands (1); which causes a cooling effect as it evaporates (1); blood vessels supplying skin capillaries dilate (1); so that more blood flows close to the surface of the skin (1).

14. Blood glucose control

1 **(a)** pancreas (1)
  (b) a hormone (1)
2 **(a)** insulin (1)
  (b) The concentration of glucose in the blood rises/increases after each meal (1). For example, after breakfast the change is from around 4 to 6 mmol per dm^3 (1).
  (c) The falls in glucose concentration would happen more slowly (1); glucose concentration would continue to rise/rise to a higher level after a meal (1).
3 **(a)** Any three from: when glucose concentration becomes low in the blood, glucagon is released (1); this breaks down stored glycogen (1); in the liver (1); to release more glucose for respiration (1).
  (b) Foods that contain high levels of sugar can quickly cause a large increase in blood glucose levels (1); and the diabetic person will be unable to produce enough insulin to reduce these levels quickly enough (1).

15. Biology six mark question 2

A basic answer would give a list of one or two comments about the two processes. There will be no attempt to compare the two processes or to reach a conclusion.

A good answer would give some advantages/disadvantages of the treatments. A final conclusion may be present, but it will be brief and may not include the information in the question.

An excellent answer would consider advantages and disadvantages for both processes, including a good description of the problems of rejection for transplants. A conclusion will be drawn that uses information from the question.

Examples of points made in the response:
Dialysis:

- Advantages – readily available; flexible, so can be administered at home or in hospital; no requirement for surgery; low cost.
- Disadvantages – does not treat the cause of kidney failure; dialysis in hospital is time-consuming (may take hours); process needs repeating regularly, so awkward if, for example, you want to go on holiday; long-term outlook is not as good as for transplant (as data in the question show).

Transplant:

- Advantages – more of a long-term solution (*a transplant is usually effective for a period of 5–10 years*); takes patients off dialysis, freeing up dialysis machines; gives people back some independence; using data in the question, this treatment gives best long-term survival rates; a single donor (if from a dead person) can treat two people who need transplant; transplant from living donor possible, as donor can live normally on one kidney.
- Disadvantages – lack of kidneys available for transplant; even when transplant is available, there are issues with rejection; recipient may not be same tissue type as the donor; recipient will need to take immunosuppressant drugs after transplant; any surgery carries other risks.

Conclusion: e.g. despite the disadvantages of possible rejection of a donor kidney, this treatment provides a better long-term solution. OR Although it seems that transplants offer a better long-term outcome, the problems with finding donors and preventing rejection means that dialysis provides a better treatment for most patients.

16. Pollution

1 pollution (1)
2 **(a)** The gas sulfur dioxide is released into the air, usually from burning fuels (1); this gas then dissolves in rain water, turning it acidic (1).
  (b) Any three from: General trend appears to be an increase in pH (1); meaning that the rain water is become less acidic (1); but the difference is small over the time period/the pattern is not very clear (1); (however) water in the lake is still acidic (1).
3 **(a)** Chemicals used in farming can dissolve into ground water (1); such as sewage/fertilisers (1).
  (b) Drawbacks – two from: land can become contaminated (1); through the use of herbicides/pesticides (1); clearing areas of tropical forest for farms can lead to global warming (1); farming of livestock release carbon dioxide and methane into the atmosphere (1). Benefits – two from: a large farm is often more efficient at food production than several small farms (1); we need farms to produce enough food for people to eat (1); use of chemicals improves yields to help alleviate food shortages (1).

17. Deforestation

1 **(a)** $32 + 34 + 3 = 69\%$ (1)
  (b) used for timber/burned for fuel (1)
  (c) Planting trees means that the rate at which carbon dioxide is taken from the atmosphere increases (1). The carbon dioxide is 'locked up' as wood (1).
2 **(a) (i)** Biodiversity is a measure of the number of different species in a habitat (1).
     (ii) Some species are destroyed in the deforestation (1); and others have their habitat destroyed (1).

(b) Decomposition of the trees by microorganisms (1); turns carbon compounds in the trees (1); into carbon dioxide in the atmosphere (1).

3 Any four of: Peat bogs are major store of carbon/store more carbon than rainforest (1); and provide an important habitat for many organisms (1); extracting peat releases carbon dioxide into the atmosphere (1); and this can contribute to global warming (1); may be a reduction in biodiversity as a result of losing peat bogs as a habitat (1).

18. Global warming

1 methane (1)

2 (a) If the temperature rose, there could be a change in the distribution of mosquitoes (1). This would mean that malaria could spread to areas outside the tropics/into Europe/into North America (1).

(b) These birds visit the UK in winter, as it is warmer than the areas they live in (1); but if they are not coming to the UK, the areas they live in must be warmer than before (1).

3 (a) photosynthesis (1)

(b) dissolves in water (1); in lakes/seas/oceans (1)

(c) Three marks for arguments (must have at least one support and one against statement) plus one mark for conclusion. Support for link: temperature rises as carbon dioxide levels rise (1); data show pattern over several years (1); data are from reliable source/same pattern seen in other sources (1). Against the link: carbon dioxide rises steadily, but temperature rise is variable (1); other factors could be affecting the temperature change as well as carbon dioxide (1). Conclusion: e.g. although the shapes of the graphs do not quite match, it is clear that the temperature rises as the level of carbon dioxide goes up (1).

19. Biofuels

1 (a) fermented (1)

(b) methane (1)

2 (a) any two from: sewage (1); animal wastes (1); plant material containing carbohydrates (1)

(b) anaerobic (1) respiration (1)

(c) The community may not have another source of fuel (1) and – in developing countries – might not be able to afford to buy fuel (1).

3 (a) The generator can only use plant/biological material (1); and most household waste will contain glass/plastic/other items that cannot be digested by bacteria (1).

(b) (i) The rate increases up to a temperature of about 35 °C (1); then decreases rapidly at higher temperatures (1).

(ii) Temperatures in the UK are below optimum temperature for biogas generation (1); so the generator needs to have efficient insulation/thick walls (1); to retain the heat generated by the respiration of the bacteria as they digest the material in the generator (1).

20. Food production

1 (a) respiration (1)

(b) (i) urine/faeces (1)

(ii) energy stored in tissues of cow = 3050 kJ − 1910 kJ − 1020 kJ (1) = 120 kJ (1)

2 Efficiency of energy transfer from crops to cow is very small (1); reducing number of steps in any food chain increases efficiency/decrease losses (1); so if humans eat crops directly, there is one fewer step and a more efficient transfer of energy (1).

3 Five marks for individual points (must have a balance of points in favour of each type) and one mark for conclusion. Indoors: graph shows that chickens lose less energy if they are kept indoors (1); this is because temperature of indoor farm can be carefully controlled (1); and movement of chickens is restricted (1); so there are fewer losses through the increased respiration as muscles work (1). Outdoors: seen by many people as more 'welfare friendly'/kinder to animals (1); chickens can find their own food, so less feed needs to be provided for them (1); disease spreads less quickly than when chickens are confined together (1); graph shows that difference between indoors and outdoors in terms of heat loss is small (1). Conclusion: for example, despite the large energy loss that occurs with free-range farming, many people prefer the chicken they eat to have been ethically farmed (1).

21. Fishing

1 (a) Prevents too many fish being caught in a particular area (1); so that breeding stocks can be maintained (1).

(b) The quota will make little difference to stocks of fish in the North Sea (1); unless other countries also adopt a quota (1).

2 (a) The proportion of farmed salmon has increased (1); from 25% to about 70%/by a factor of nearly three times (1); OR the proportion of wild salmon has decreased (1); from about 70% to 25%/by a factor of one-third (1).

(b) Farmed salmon are kept in cages, so do not have as much room to swim/to expend energy (1); as they are provided with food, they may overeat/do not exercise while swimming to find their own food (1).

(c) Fewer wild salmon are removed from the sea, so numbers go up (1); this may have an effect on other fish lower down the food chain on which salmon feed (1); and also could lead to scarcity of resources for larger population of wild salmon (1).

3 Mesh size of net is important (1); only fish of a certain size are able to breed (1); nets need to catch fish larger than this (1); so that a number of breeding fish are left in the sea (1).

22. Sustainable food

1 (a) glucose syrup (1)

(b) The air contains oxygen (1); which the fungi need to respire (1).

2 (a) water (1)

(b) The fungi can be grown quickly and cheaply (1); as the fungi are fed directly on glucose/there are few steps in the food chain, the production of mycoprotein is efficient in terms of energy transfer (1).

3 Three marks for arguments (must have at least one advantage and disadvantage statement) and one mark for conclusion. Advantages: more choice for customers on what they eat and when (1); previously useless desert land is being used to provide food for people (1). Disadvantages: food is flown in from Israel/food has increased amount of 'food miles' (1); this is potentially bad for the environment (1); land converted from desert is used to provide choice for Western shoppers rather than food for poorer countries (1); desert habitat is destroyed (1). Conclusion: for example, although it is nice for us to have choice in our food, eating food with high 'food miles' is bad for the environment as it involves increased carbon dioxide emissions (1).

4 The respiration of the fungi inside the fermenter releases energy (1); so the fermenter needs to be kept cool so that it does not overheat (1); otherwise the enzymes controlling respiration would denature (1).

23. Biology six mark question 3

A basic answer would give some information about how deforestation leads to more carbon dioxide in the atmosphere. However, the answer is likely to be a series of unconnected statements.

A good answer would link points together more, but is likely to concentrate on the idea of deforestation leading to fewer trees taking in carbon dioxide and the burning of the wood to release carbon dioxide. There may be isolated information about what the land is subsequently used for and how this contributes to climate change.

An excellent answer would give a full answer, and would consider aspects such as the use of the cleared land, and how farming/biofuel production will affect climate change. In addition, there is likely to be better explanation of how the data in the graph only presents part of the story.

Examples of points made in the response:

- Fewer trees will lead to less carbon dioxide removed from the atmosphere through photosynthesis.
- Trees may be burned for fuel.
- Burning the trees releases the carbon stored in their biomass as carbon dioxide.
- Carbon dioxide adds to global warming.
- This has the effect of increasing global atmospheric temperature.
- Graph shows decrease in rate of deforestation in Brazil.
- But deforestation is occurring elsewhere in addition.
- Climate change is a long-term process, so even though deforestation has slowed in Brazil overall over the period this is too short a timescale to see an effect on climate/ there are too many fluctuations in the data.
- Land cleared may be used for farming.
- Farming animals e.g. cattle releases other gases that contribute to global warming.
- Methane is a common gas released through farming activities.
- Land may be used for producing biofuels.
- Although these are less polluting than fossil fuels, they are not carbon neutral because of production and distribution.

Chemistry answers

24. The early periodic table

1 the elements' atomic weight (1)
2 **(a)** In Newlands's table the groups are arranged in horizontal rows (1). In the modern periodic table the groups are in vertical columns (1).
  (b) noble gases (1)
  (c) Li and Be (1)
  (d) In the group starting with lithium (Li/2) the law of octaves works up to potassium (K/16) (1) and also for rubidium (Rb/30) and caesium (Cs/44) (1). The elements copper (Cu/23), silver (Ag/37) and thallium (Tl/53) don't fit as alkali metals (1).
3 **(a)** increasing atomic weight (mass) order (1)
  (b) (i) gallium and germanium (1)
    (ii) They hadn't been discovered (1).

25. The modern table

1 **(a)** A and F (1)   **(b)** D (1)
  (c) two of C, D and E (1)

2 **(a)** In the modern periodic table elements are arranged in order of increasing number of protons (1), with elements placed in the same group if they have the same number of electrons in their highest energy level (1). Elements in the same row (period) have the same number of occupied energy levels (1).
  (b) (i) The main group number of the element is the same as the number of electrons in the outer shell (1).
    (ii) The exception is in Group 0 (or 18, noble gases) (1).
3 **(a)** To make the elements fit into groups with similar properties, Mendeleev moved the position of some elements (1) and left gaps for elements yet to be discovered (1).
  (b) It was not accepted because there was no explanation/understanding of the idea (1).
  (c) We can make predictions as elements in the same group usually have similar properties (1), and as you go down a group there is a general trend in properties/as you move along a period (row) certain properties change gradually (1).

26. Group 1

1 **(a)** rubidium (1)   **(b)** rubidium (1)
2 **(a)** Alkali metal compounds generally have ionic bonds (1), are soluble in water (1) and are white or colourless (1).
  (b) When all alkali metals react with water they produce metal hydroxides (1), which dissolve in water (1) to form alkaline solutions (1).
3 **(a)** sodium + water → sodium hydroxide (1) + water (1)
  (b) (i) $2Na + Cl_2$ (1) → $2NaCl$ (1)
    (ii) $4Na + O_2$ (1) → $2Na_2O$ (1)
  (c) Because it is very reactive (1), it is stored under oil to stop it reacting with the air/moisture in the air (1).
4 This was not a safe way of disposing of sodium as sodium is less dense than water (1), so the tin will float (1), and sodium is very reactive (1), so will react fiercely/explode in water (1).

27. Transition metals

1 In order, top to bottom: false (1); true (1); false (1); false (1).
2 **(a)** D and E (1)   **(b)** A (1)   **(c)** B (1)   **(d)** E (1)
3 **(a)** The transition elements are found between calcium and zinc (1), and between yttrium and cadmium (1).
  (b) Alkali metals generally have lower melting points than transition metals (1).
4 **(a)** Iron(II) chloride is $FeCl_2$ (1) and iron(III) chloride is $FeCl_3$ (1).
  (b) NaCl (1)
  (c) Iron can form more than one compound as it can form ions with different charges (1). The alkali metals only form the one compound as they only form one ion (1).

28. Group 7

1 As you go down the group, the reactivity decreases (1), and melting and boiling points increase (1).
2 **(a) (i)** All halogens are poisonous/dangerous (1).
    (ii) Fluorine would be too reactive (1).
  (b) Iron(III) chloride (1) and iron(II) bromide (1). *You need the (III) and (II) in the answers to get credit.*
  (c) $Fe + I_2$ (1) → FeI_2 (1)
  (d) As you go down the Group 1 elements more electron energy levels are occupied (1). The reactivity of the Group 7 elements decreases down the group as it is harder to gain an electron (1) the higher the energy level of the outer electrons (1).

3 (a) $Cl_2(g) + 2NaBr(aq) \rightarrow 2NaCl(aq)$ (1) $+ Br_2(aq)$ (1)
  (b) displacement
  (c) Halogens higher up the periodic table (more reactive halogens) will react with halide (halogen) ions further down the periodic table (less reactive halide ions) (1). OR A more reactive halogen will displace a less reactive halogen from a compound (1).

29. Hard and soft water

1 magnesium (1) and calcium (1)
2 (a) Soap forms a lather in soft water (1) but forms a scum in hard water (1).
  (b) Hard water contains calcium or magnesium ions (1); soft water does not contain these ions (1).
3 (a) Measure a volume of water (with measuring cylinder) and place in conical flask (1). Add soap solution from burette, to hard water, 1 cm³ at a time (1). Shake flask after each addition and note volume of soap solution needed to produce a permanent lather (bubbles) (1). The greater the volume of soap solution needed, the harder the water (1).
  (b) Any two from: volume of water (1); concentration/type of soap solution (1); amount of shaking (1).
4 (a) from the rocks it runs through (1)
  (b) Heating breaks down the hydrogencarbonate ions to form carbonate ions (1). The carbonate ions then react with the calcium and magnesium ions (1), which form a precipitate (1).

30. Softening hard water

1 Calcium and/or magnesium ions (1).
2 (a) It forms a scum (1) with the calcium/magnesium ions (1) in the hard water, so more soap is needed (1).
  (b) insoluble compounds of calcium/magnesium OR calcium/magnesium carbonate/insoluble solids formed from the ions in the hard water (1)
  (c) The calcium in hard water helps build strong bones and teeth (1). It also helps reduce heart disease (1).
3 (a) The compounds do not break down on heating (1) and the calcium/magnesium ions remain in solution (no precipitate is formed) (1).
  (b) (i) The calcium ions are removed (1) from the solution as a precipitate (1) of solid calcium carbonate (1).
    (ii) $MgSO_4(aq) + Na_2CO_3(aq)$ (1) $\rightarrow MgCO_3(s) + Na_2SO_4(aq)$ (1)
  (c) The calcium/magnesium ions go into the resin (1) and are replaced by hydrogen/sodium ions (1).

31. Purifying water

1 (a) The water may be polluted with chemicals that cannot be removed by treatment (1).
  (b) (i) $Al_2(SO_4)_3$ (1)
    (ii) In the sediment tank the lumps of solid will settle/sink (1) to the bottom of the tank.
  (c) Solid particles are removed (1).
  (d) To kill bacteria (1) that are not removed by the filter beds (1).
  (e) Advantage: fluoride helps prevent tooth decay (1). Disadvantage: some people may react/be allergic to fluoride (1).
2 (a) The water is heated to (boil) evaporate it, then it is condensed (back to liquid) (1). All dissolved solids stay in the original water (are left behind) (1).
  (b) Distillation needs heat energy (1), which is expensive (1).

  (c) Substances that evaporate easily/boil at a lower temperature than water (1).

32. Chemistry six mark question 1

A basic answer will have a simple description of some of the chemical reactions or trends in reactivity for each group of elements – the alkali metals and the halogens – or a description of reactions and trends for only one of the two groups.

A good answer will have a clear description of the chemical reactions and the trends in reactivity for both groups of elements.

An excellent answer would include a clear, detailed and balanced explanation of the different trends in chemical reactions found in both groups. The explanation would include reference to the loss or gain of outer electrons.

Examples of points made in the response:
- The alkali metals react with non-metals to form ionic compounds.
- e.g. $2Na + Cl_2 \rightarrow 2NaCl$.
- The alkali metals react with water to release hydrogen gas and produce an alkali/hydroxide.
- e.g. $2Na + 2H_2O \rightarrow 2NaOH + H_2$.
- The halogens react with metals to form ionic compounds.
- $F_2 + 2Li \rightarrow 2LiF$.
- The reactivity of alkali metals increases down the group.
- Because the highest energy level/outer shell is further from the nucleus.
- So electrons are lost more easily.
- The reactivity of halogens decreases down the group.
- Because the highest energy level/outer shell is further from the nucleus.
- Which makes it harder for electrons to be gained.
- A more reactive halogen will displace a less reactive halogen from a solution of its salt.

33. Calorimetry

1 Missing measurements are: mass of burner + fuel at the end (1); temperature at start (1); and temperature at the end (1).
2 (a) (i) Heat energy lost to the surroundings (1).
    (ii) Reduce heat energy losses (1) by using insulation/draught excluders around experiment (1).
  (b) Use the results to work out the temperature rise per gram of fuel burned (1). The fuel that produces the biggest temperature rise gives out most heat energy (1).
3 (a) $\Delta T = 22 - 20 = 2\,°C$ (1)
    $Q = C \times m \times \Delta T = 4.2 \times 100 \times 2$ (1) $= 840$ (1) joules
  (b) Repeating and averaging shows up errors to produce more accurate results (1).
  (c) $4\,°C$ (1)

34. Energy level diagrams

1 (a) 2 (1)   **(b)** 3 (1)
2 (a) The products have less energy than the reactants (1) so energy has been given out (1).
  (b) 50 kJ (1)
3 (a) 30 kJ (1)   **(b)** 80 kJ (1)
  (c) The energy needed to break bonds (1) is more than the energy released when bonds are formed (1) so the reaction takes in heat overall (1) and is therefore endothermic.

(d)

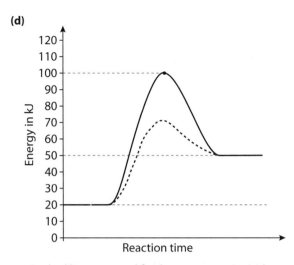

Dashed line starts and finishes at same point (1) but peak is below first graph (1).

35. Bond energies

1 **(a)** Energy in: $1 \times N\equiv N = 1 \times 941\,kJ = 941\,kJ$;
$3 \times H-H = 3 \times 436\,kJ = 1308\,kJ$;
total energy in $= 941\,kJ + 1308\,kJ = 2249$ (1) kJ.
Energy out $= 6 \times N-H = 6 \times 391\,kJ = 2346$ (1) kJ.
Energy change $= 2346\,kJ - 2249\,kJ = 97\,kJ$ (1).
  (b) The reaction is exothermic (1) as more energy is given out than taken in (1).
2 The energy to break bonds (and start the reaction) is very high (1) so energy has to be put in at the start (1).
3 **(a)** Activation energy (energy in to break bonds):
$1 \times H-H = 1 \times 436 = 436\,kJ$;
$1 \times Br-Br = 1 \times 193 = 193$ (1) kJ. Total $= 629\,kJ$ (1).
  (b) ($2 \times H-Br$ bond energy) $-629\,kJ = 103\,kJ$ (1);
($2 \times H-Br$ bond energy) $= 103 + 629 = 732\,kJ$;
$H-Br$ bond energy $= 732/2 = 366\,kJ$ (1)

36. Hydrogen as a fuel

1 **(a)** Hydrogen can be burned as a fuel (1) or it can be used in a fuel cell to make electricity (1).
  (b) When hydrogen is used as a fuel the only product is water (1), which is harmless/no effect on our environment (1). When diesel or petrol burn the products can affect (pollute) our environment (1).
2 **(a)** $2H_2 + O_2$ (1) $\rightarrow 2H_2O$ (1)
  (b) $CH_4 + H_2O$ (1) $\rightarrow CO + 3H_2$ (1)
  (c) oxygen (1)
  (d) Method 1 uses up fossil fuel reserves, whereas method 2 uses a renewable resource (1). Method 1 produces poisonous or polluting products, but method 2 does not produce poisonous or polluting products (1).
  (e) Most electricity is made by burning fossil fuels, so it uses up a finite resource (1) and it causes pollution (1).
3 **(a)** oxygen (1)
  (b) Any two from: we need to develop a sustainable way of producing hydrogen (1); we need to develop cheaper/more efficient fuel cells (1); we need to develop a method of storing hydrogen safely (1).

37. Tests for metal ions

1 **(a)** lithium (1)   **(b)** solid (1)
2 **(a)** Dip a clean (metal) rod in the compound being tested and hold it in a pale blue Bunsen burner flame (1). The colour of the flame can tell us which metal might be present (1).

  (b) Ionic substances contain two ions (1) and each of the ions reacts differently (1).
  (c) carbon dioxide (1)
  (d) calcium (1) carbonate (1)
3 **(a)** precipitation (1)
  (b) The missing information is: Cu^{2+}(aq) (1); green (1); brown (1).
  (c) When excess sodium hydroxide is added (1) the aluminium precipitate will re-dissolve (1).
  (d) $2\,NaOH(aq) + Ca(NO_3)_2(aq) \rightarrow Ca(OH)_2$ (1) (s) (1) $+$ $2NaNO_3$ (aq) (1)

38. More tests for ions

1 First add hydrochloric acid; if carbonate ions are present a gas is produced that will turn limewater cloudy (1). To test for sulfate ions, add barium chloride and hydrochloric acid; if sulfate ions are present then a white precipitate will be formed (1).
2 **(a)** Add silver nitrate and nitric acid to each (1). Sodium bromide forms a cream precipitate and sodium chloride forms a white precipitate (1).
  (b) Add barium chloride and hydrochloric acid to both. Sodium sulfate forms a white precipitate and sodium chloride does not form a precipitate.
3 **(a)** Test 1 shows that certain metals like sodium are not present (1). Test 2 shows that aluminium ions are present (1).
  (b) Test 3 shows that there are no carbonate ions present (1). Test 4 shows that iodide ions are present (1).
  (c) Test water samples from different parts of the river working downstream (1). The factory just upstream from the first positive test will be producing the pollution (1).

39. Titration

1 titration (1); indicator (1)
2 **(a)** volume $= 15\,cm^3 \div 1000 = 0.015\,dm^3$; moles of LiOH $= 0.015 \times 0.2 = 0.003$; 1 mole of LiOH neutralises 1 mole of HNO_3 so there are 0.003 moles of HNO_3 (1); $N = C \times V$, $0.003 = C \times 0.005$ (1) so $C = 0.003/0.005 = 0.6\,mol/dm^3$ (1)
  (b) 1 mole LiOH $= (7 + 16 + 1) = 24\,g$ (1); $0.2\,mol/dm^3$ $= 0.2 \times 24 = 4.8$; concentration $= 4.8\,g/dm^3$ (1)
3 **(a)** volume $= 7.5\,cm^3 \div 1000 = 0.0075\,dm^3$; moles of HCl $= 0.0075 \times 1.0 = 0.0075$; 1 mole of HCl neutralises 1 mole of NaOH so 0.0075 moles of NaOH (1); $N = C \times V \therefore 0.0075 = C \times 0.025$ (1) $C = 0.0075/0.025 = 0.3\,mol/dm^3$ (1)
  (b) 1 mole HCl $= (1 + 35.5) = 36.5\,g$ (1) so $1.0\,mol/dm^3$ $= 36.5\,g/dm^3$ (1)

40. Chemistry six mark question 2

A basic answer will have a simple description of the properties of each of the three types of water and the problems caused by hard water.

A good answer will have a clear description of the properties of each type of water, the chemicals in each and some discussion of advantages and disadvantages.

An excellent answer will have a detailed description of the properties of each type of water, the chemicals in each, and advantages and disadvantages, including any chemical reactions where relevant.

Examples of points made in the response:

- Permanent hard water contains calcium and magnesium ions.
- Temporary hard water contains hydrogen carbonate ions as well as calcium and magnesium ions.
- Boiling temporary hard water removes the hardness.
- The hydrogen carbonate ions break down to form carbonate ions.
- Which react with the calcium and magnesium ions to form a precipitate (calcium carbonate or magnesium carbonate).
- Disadvantages of permanent and temporary hard water: they form an insoluble scum with soap/they form a solid scale in kettles and boilers/they cause extra expense to remove ions if soft water is required.
- Advantage of permanent and temporary hard water: the calcium ions help build strong bones and teeth.
- Soft water does not contain calcium or magnesium ions.
- Advantages of soft water: forms lather easily with soap/does not need expensive treatment to remove ions before use.
- Disadvantage: no added health benefits.

41. The Haber process

1 **(a)** Nitrogen is obtained from the air (1) and hydrogen is obtained from methane and steam (1).
 (b) Iron (1) is present and it acts as a catalyst (1).
2 **(a)** The reaction is reversible/comes to equilibrium (1).
 (b) The unreacted nitrogen and hydrogen are recycled (1).
3 **(a)** $N_2(g) + 3H_2(g)$ (1) $\rightleftharpoons 2NH_3(g)$ (1)
 (b) The reaction is reversible/an equilibrium/goes both ways (1).
4 **(a)** An advantage is a higher yield of ammonia (1); a disadvantage is that it is expensive to maintain (1).
 (b) (i) A lower temperature produces a higher yield of ammonia (1).
 (ii) The reaction might be too slow (1).

42. Equilibrium

1 **(a) (i)** A reaction that takes in heat/energy (1).
 (ii) A reaction that can go both ways (1).
 (b) If the temperature is raised the yield in the endothermic direction will increase (1), so there will be more NO_2 (1), so the colour will get darker/more brown (1).
 (c) If the gas pressure increases, the yield in the direction that forms the fewest gas molecules will increase (1), so there will be less NO_2/more N_2O_4 (1), so the colour will get lighter/more yellow (1).
2 **(a)** Decreasing the pressure has no effect on the amount of product (1) as there are the same number of gas molecules on each side (1).
 (b) Decreasing the pressure would produce less product (1) as the equilibrium position moves in the direction that produces the most gas molecules (1).
3 A low temperature produces more yield in the exothermic direction (1) and a high pressure produces more yield in the direction that forms the fewest gas molecules (1).

43. Alcohols

1 **(a)** OH group (1)
 (b) $C_4H_9OH/C_4H_{10}O$ (1)
2 **(a)** carbon dioxide (1) and water (1)
 (b) Butanol could be used as a solvent (1) or as a fuel (1).

3 **(a)** CH_3OH/CH_4O (1) and

H—C—C—C—O—H or H—C—C—C—H (1).

 (b) sodium (1); hydrogen (1)
4 **(a)** Fermentation produces ethanol from plant sugar (1), which is renewable as the plants can be grown again (1) each year.
 (b) $C_2H_5OH + 3O_2 \rightarrow 2CO_2 + 3H_2O$ (1 mark for both formulae correct and 1 mark for correct balancing.)
 (c) Health problem: heart disease/liver damage (1); social problem: increased aggression/families short of money (1).
 (d) oxidation (1)   **(e)** ethanoic acid (1)

44. Carboxylic acids

1 **(a)** C_3H_7COOH or $C_4H_8O_2$ (1)
 (b) $C_nH_{2n}O_2$ (1)
2 **(a)** butanol (1)
 (b) C_2H_5COOH or $C_3H_6O_2$ (1)

H—C—C—C (1)

 (c) carbon dioxide (1)
 (d) Carboxylic acids react with alcohols (1) to form esters (1). OR Carboxylic acids react with metals (1) to form hydrogen (1).
3 **(a)** Missing information: CH_3COOH (1); 3 to 4 (1); any value less than 6 minutes (1).
 (b) The $\rightleftharpoons$ sign means that in a weak acid not all the acid molecules break up into ions/some of the ions combine to form acid again (1). In a strong acid all the acid molecules break up into ions (1).
 (c) The difference is because the HCl breaks up completely into ions and the ethanoic acid does not (1). This means there are more H^+ ions in the HCl than CH_3COOH (1) so the HCl has a lower pH (1).

45. Esters

1 A carboxylic acid (1) and an alcohol (1)
2 **(a)** ethanoic acid (1) and ethanol (1)
 (b) It acts as a catalyst (1).
 (c) The reactants (and products) in this reaction can be flammable (1) so a water bath is less likely to start a fire (1).
 (d) (i) the excess acid (1)   **(ii)** a sweet smell (1)
3 **(a)** Ethyl ethanoate (1), and it is an ester (1).

 (b)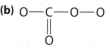

 (c) as a flavouring (1) or in perfumes (1)

46. Using organic chemicals

1 **(a)** Missing information: $C_3H_5O_2$ (1), ester (1), carboxylic acid (1).
 (b) (i) compound A (1)   **(ii)** compound C (1)
2 ethyl ethanoate, any two from: perfume, food flavouring or solvent (1); ethanol: fuel and solvent (1)

3 **(a)** $2CH_4 + 2O_2$ (1) $\rightarrow CH_3COOH + 2H_2O$;
C_2H_5OH (1) $+ O_2 \rightarrow CH_3COOH + H_2O$

(b) Method 1 is a faster reaction/is a continuous process (1) and so more ethanoic acid will be produced in the same time (1).

(c) The main advantage of using ethanol from fermentation is that it is formed from a plant/renewable source (1) and does not use up reserves of fossil fuels (1).

47. Chemistry six mark question 3

A basic answer will have a simple description of how the apparatus can be used to find the volume of sodium hydroxide solution needed to neutralise a set volume of each of the vinegars.

A good answer will have a clear description of how the volume of sodium hydroxide solution needed to neutralise a set volume of each of the vinegars can be found by experiment, including information on how to make the test fair and how to use the results.

An excellent answer will have a detailed description of the experimental procedure to compare the acid content in the vinegars by titration. It will also include details of the variables that need to be controlled, how the acid concentration is compared and how the results can be made more reliable.

Examples of points made in the response:

- Vinegar contains an acid that can be neutralised by sodium hydroxide solution.
- Set volumes of vinegar should be measured using the measuring cylinder.
- And placed in the beaker along with the indicator.
- The sodium hydroxide solution should be added using the burette.
- Until the indicator changes colour.
- The larger the volume of sodium hydroxide solution, the higher the concentration of acid in the vinegar.
- To make the test fair use the same volume of vinegar and the same sodium hydroxide solution each time.
- The results of the titration can be made more reliable by repeating the titration and averaging the results for each acid.
- As acids and alkalis are used, safety goggles should be worn at all times.

Physics answers

48. X-rays

1 A (1)

2

	Transmit X-rays	Absorb X-rays
Bones and metals		✓ (1)
Soft tissues	✓ (1)	

3 Any three from: X-ray sources should be shielded by lead to stop X-rays from being emitted in all directions (1); warning lights/signs to show when X-ray machines are operating (1); operators retreat to safe areas (1); lead aprons worn (1); patients should not have X-ray procedures often (1); use of film badges to monitor exposure (1).

4 It was discovered that X-rays are harmful/ionising (1) so exposure should be kept as low as possible (1).

5 Metals absorb X-rays completely (1); bones absorb most X-rays but do transmit some of them (1), and soft tissue absorbs a few X-rays and transmits most of them (1).

49. Ultrasound

1 **(a)** The sound wave is reflected (1) from a boundary or surface between different tissues or the surface of the foetus (1).

(b) distance travelled by sound $= 1500$ m/s $\times 0.00008$ s (1) $= 0.12$ m (1); so distance to foetus $= \frac{1}{2} \times 0.12$ m $= 0.06$ m (or 6 cm) (1)

2 **(a)** time between peaks $= 15 \times 10^{-6}$ s (1); time taken for ultrasound to travel one way through skull $= 15 \times 10^{-6}/2 = 7.5 \times 10^{-6}$ s (1)

(b) rearranging gives $v = s/t$ (1); 0.12 m/8 $\times 10^{-6}$ s (1) $= 1.5 \times 10^3$ or 1500 m/s (1)

50. Medical physics

1

Medical method	Uses X-rays	Uses ultrasound
scanning pregnant women		✓ (1)
killing cancer cells in patients	✓ (1)	
obtaining pictures of broken bones	✓ (1)	
breaking up kidney stones		✓ (1)

2 X-rays are ionising radiation, ultrasound is not (1).

3 In a CT scan, X-rays (1) pass through the body. The source and detector, called a charge-coupled device (1), rotate around the subject. An image is formed by combining signals from many directions (1).

4 X-rays – advantages: they reveal injuries/disease that cannot be detected by other methods (1); disadvantages: ionising radiation and can harm cells/cause cancers (1). Ultrasound – advantages: it is not dangerous so can be used with pregnant women (1); disadvantages: relatively poor resolution of images (1).

51. Refraction in lenses

1 refraction (1); towards (1)

2 **(a)** Light is refracted so that it converges to/meets at/passes through (1) a point.

(b) (i) the principal focus (1)　**(ii)** the focal length (1)

3 refractive index $= \sin i/\sin r = \sin 60°/\sin 33°$ (1) $= 0.866/0.545$ (1) $= 1.59$ (1)

4 $\sin i =$ refractive index $\times \sin r = 1.33 \times \sin 30°$ (1) $= 1.33 \times 0.5$ (1) $= 0.665$ (1); $i = 41.7°$ (1)

52. Images and ray diagrams

1 inverted (1); magnified (1); real (1)

2 **(a)** Travels parallel to the axis to the lens (1) and is refracted to pass through the principal focus (1).

(b) real (1); inverted (1); magnified (1)

(c) A real image can be displayed on a screen (1) and is formed when light rays converge (1). A virtual image cannot be displayed on a screen (1), and rays of light appear to come from a point behind a lens (1).

3 All 3 rays completed correctly (1), rays meeting at point (1), image drawn 4.5 cm from lens (1).

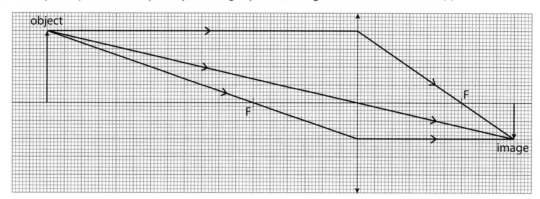

53. Real images in lenses

1 magnification = image height/object height = 70 cm/3.5 cm (1) = 20 (1)
2 20 cm – real, inverted, same size (1); 14 cm – real, inverted, magnified (1); 22 cm – real, inverted, smaller (1)
3 (a) magnification = image height/object height = 10 cm/2.5 cm (1) = 4 (1)
  (b) magnification = image height/object height = 2 cm/2.5 cm (1) = 0.8 (1)
4 Ray diagram similar to that shown. All three rays drawn correctly (1); image at 7.5 cm from lens (1); image 1.5 times taller than object (1). *The diagram here has an object 1 cm high. It doesn't matter how tall you have made your object, as long as the image is 1.5 times as tall.* Magnification = 1.5 cm/1 cm (1) = 1.5 (1). *If your object was a different height, you get both marks if you correctly calculated that the magnification was 1.5 times the height.*

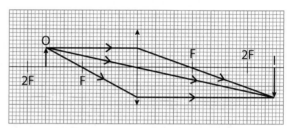

54. Virtual images in lenses

1 (a) The dashed line shows where the rays seem to have come from (1) although they have not really travelled along these lines (1).
  (b) upright/erect (1); bigger/magnified (1); virtual (1)
  (c) magnification = 7 cm/2 cm (1) = 3.5 (1)
2 Ray that would have passed through focus refracted parallel to axis (1); ray that travels parallel to axis refracted so that it would have passed through focus (1); rays meet at point on image (1); distance of image to lens measured and scaled = 10 cm (1).

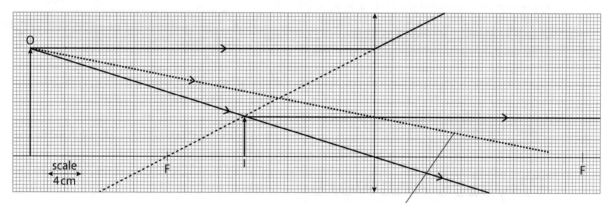

This is not a ray, but it will help you to draw accurate diagrams if you draw the full line in faintly before showing what actually happens to the light.

55. The eye

1 A: retina
  B: lens
  C: pupil
  D: iris
  E: ciliary muscles.
  All labelling correct (3); 2–3 correct (2); 1–2 correct (1).
2 (a) cornea, lens (1) *Both needed.*
  (b) iris/pupil (1)
  (c) ciliary muscles (1)
3 They may find it difficult to see in dim light (1) if the pupil does not let enough light through (1). They may be dazzled in bright light (1) if the pupil lets too much light through (1).

56. Range of vision

1 C (1)

2 There is no limit to the distance of an object from the eye (1) for which a sharp image can be formed on the retina (1).

3 (a) The amount of light let in is controlled by the iris/pupil in the eye and by the aperture in a camera (1). They both have a convex lens to help focus the light (1).

(b) Any two from: in the eye the shape of the lens changes, but in a camera the position of the lens changes (1); cameras have a shutter that only allows light in while the picture is being taken, the eye does not have anything similar (1); the camera records the image on film or using a charge-coupled device, but the eye converts light to nerve signals (1).

4 The ciliary muscles relax (1) and the lens becomes thinner/flatter (1).

57. Correction of sight problems

1 A: short sight (1)
B: long sight (1)
C: long sight (1)
D: short sight (1)

2 (a) The converging (1) lens refracts the light so that a sharp image of an object that is near (1) to the eye is formed (on the retina).

(b) Any two points from: contact lenses are less likely to be knocked off (1); contact lenses cover the whole field of view (1); no frames to obscure part of the vision (1); contact lenses don't get steamed up (1); contact lenses give the person a natural appearance (1).

3 Concave lens makes rays diverge (1); cornea and lens in eye make rays converge (1), to meet on retina (1).

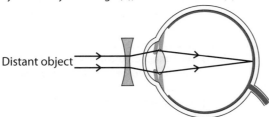

Distant object

58. Power of a lens

1 B (1); C (1)

2 (a) -20 cm is -0.20 m (1); $P = 1/f = 1/-0.20$ m (1) $= -5$ (1) dioptres

(b) diverging/concave (1)

3 (a) $f = 1/P = 1/2.5$ dioptres (1) $= 0.40$ (1) m (1)

(b) Any four from: same convergence can be achieved with a less curved lens (1) so the lens will be lighter (1) and so more comfortable to wear (1), and thinner (1) so will look better (1).

59. Total internal reflection

1 (a) A: straight line drawn in air from point of incidence (1) at angle greater than angle of incidence (1).

(b) B: straight line drawn in glass from point of incidence (1) with angle of reflection equal to angle of incidence (1).

2 Light from the object on the bottom hits the surface at an angle greater than the critical angle (1) and is reflected to the diver (1). The light appears to come from a point above the surface (1).

3 (a) refractive index $= 1/\sin c = 1/\sin 39.12\,°$ (1) $= 1/0.6309$ (1) $= 1.585$ (2) *Both 1.59 and 1.5849 are awarded only 1 mark as the appropriate degree of accuracy is 4 significant figures because the critical angle is given to 4 significant figures.*

(b) $\sin c = 1/$refractive index $= 1/2.4; = 0.417$ (1); $c = \sin^{-1} 0.417 = 24.6\,°$ (1) *25° is acceptable.*

60. Other uses of light

1 The ray is reflected off the sides of fibre until it reaches the end (1); law of reflection obeyed (approximately) at each point (1). *Diagram may not look exactly like the example shown due to small errors in angles.*

2 (a) The light strikes the boundary between the glass and air at angles greater than the critical angle (1) for the glass and undergoes total internal reflection (1) at each point.

(b) Light is carried into the stomach by one (bundle of) optical fibres (1). Light reflects off the stomach lining and is carried by another bundle of optical fibres (1) to form an image in the doctor's eye.

3 (a) Any two points from: lasers produce a very narrow (1), powerful (1) beam of light (1) that can cut/cauterise/burn materials (1).

(b) The laser is used to cut the cornea (1) to change its curvature (1). The surgery makes the lens less curved (1) so that light rays from distant objects do not converge in front of the retina (1).

61. Physics six mark question 1

A basic answer will give a brief description of a cause of long or short sight and a method of how it is corrected.

A good answer will include a description of causes of long and short sight and how they are corrected using lenses.

An excellent answer will include a clear and detailed description of the causes of long and short sight and an explanation of how they are corrected using converging and diverging lenses and laser surgery.

Examples of points made in the response:
- Light from an object is refracted/converged
- by the cornea
- and the lens
- to form an image on the retina.
- In short sight: the image forms in front of the retina
- because the eyeball is too long
- or the cornea/lens has too sharp a curve.
- In long sight: the image forms behind the retina
- because the eyeball is too short
- or the cornea/lens is not curved enough.
- Spectacles and contact lenses make the image form on the retina
- for short sight a diverging lens is needed
- for long sight a converging lens is needed.
- Contact lenses are placed directly on the cornea.
- Laser surgery changes the shape of the cornea
- by making precise incisions/cuts in the cornea.

62. Centre of mass

1 E, B, C, A, D (all correct, 2 marks; 3 or 4 in correct order, 1 mark)

2 The centre of mass is a point in an object (1) at which the mass may be thought to be concentrated (1).

3 (a) 8 and 0 (both needed for 1 mark)
(b) The centre of mass is vertically below the point from which the digit hangs (1). Only the digits shown with a vertical line of symmetry will hang correctly (1).

63. The pendulum

1 B (1)
2 (a) (i) The measurement of the frequency or period will be more accurate (1) because an error in the reading is spread over more swings/an error will have a bigger effect on just one swing (1).
(ii) frequency = number of swings/time taken = 24/60 (1) = 0.4 (1) Hz (1)
(b) $T = 1/f = 1/0.4$ Hz (1) = 2.5 s (1) If you used an incorrect answer from question **2(a)(ii)** correctly in the calculation you can give yourself the marks.
(c) Change the length of the rope (1).
3 $f = 1/T = 1/0.75$ s (1) = 1.33 (1) Hz

64. Turning effect and levers

1 (a) moment = force × perpendicular distance = 200 N × 1.5 m (1) = 300 Nm (1)
(b) A (1), D (1)
2 $M = F \times d$; clockwise moment = 200 N × 1.8 m = 360 Nm (1); anticlockwise moment = 150 N × 2.4 m = 360 Nm (1); the seesaw is balanced because clockwise moment = anticlockwise moment (1).
3 (a) The force acting on the rock/the load is greater than the force that the workman uses/the effort (1) because the pivot is closer to the rock/load than the man/effort (1).
(b) $M = F \times d$; when the rock is just about to move, anticlockwise moment = clockwise moment. 200 N × 1.5 m = W × 0.2 m (1); W = 300 Nm/0.2 m (1) = 1500 N (1).

65. Moments and balance

1 If the seesaw is balanced then clockwise moment about the pivot = anticlockwise moment about the pivot; clockwise moment = $M = F \times d = 600$ N × d m (1); anticlockwise moment = 150 N × 1.6 m (1) = 240 Nm. It is balanced so 600 N × d m = 240 Nm (1); d = 240 Nm/600 N = 0.4 m (1).
2 The maximum force applied is just when the moments are balanced (1); 400 N × 1 m = F × 0. 2 m (1); F = 400 Nm/0.2 m = 2000 N (1).
3 The weight of the plank acts at its centre of mass/midpoint, and that is 1 m horizontally from where the end of the plank rests on the ground (1); anticlockwise moment of weight of plank about the end touching the ground = 180 N × 1 m = 180 Nm; clockwise moment of the reaction of the fence on the top end of the plank = F × 2 m; the plank is balanced so 180 Nm = F × 2 m (1); F = 180 Nm/2 m = 90 N (1), which is less than the 100 N needed to collapse the fence.

66. Stability

1 (a) C, A, B (1)
(b) C has a wider base than A and B so is most stable (1). B has a higher centre of mass than A so is least stable (1).
2 (a) The stools should have a wide base (1) and a low centre of mass (1).
(b) Only when the stool is tilted at a large angle does the line of action of the weight (1) fall outside the base of the stool (1), causing it to topple.

3 When the toy is pushed over there is a (perpendicular) distance between line of action of its weight and the pivot/the point where it touches the ground (1), and this causes a moment that turns the toy upright again (1). Sketch similar to the one below showing this (1).

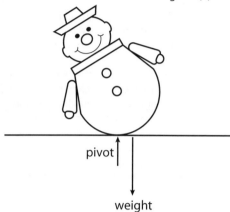

67. Hydraulics

1 compressed (1); transmitted (1); exerted (1)
2 The effort acting over the area of the small piston produces a pressure (1). This is transmitted to the load piston/equally in all directions (1). The load is bigger than the effort because the area of the load piston is larger (1).
3 (a) $P = F/A = 3.2 \times 10^4$ N/8.0×10^{-2} m² (1) = 4.0×10^5 (1) N/m² or Pa (1)
(b) $F = P \times a = 4.0 \times 10^5$ Pa × 4.0×10^{-4} m² (1) = 160 (1) N (or 1.6×10^2 N)

68. Circular motion

1 (a) centripetal (1)
(b) (i) gravity (1) **(ii)** friction (1) **(iii)** tension (1)
2 When it is moving in a circle the direction of the weight is constantly changing (1) so its velocity is changing/it is accelerating (1). A force is needed to make an object accelerate/change velocity (1).
3 (a) The runner on the inside track needs more friction/centripetal force (1) because the radius of the circle is smaller (1).
(b) At high speed there may not be sufficient friction (1) between the tyres and the road to make the vehicle turn because the centripetal force required increases with speed (1).
(c) The centripetal force required to make the lorry turn increases (1) with the increased mass of the load (1).

69. Physics six mark question 2

A basic answer will describe the turning effect of the load on the crane and suggest a way of making the crane more stable.

A good answer will describe the moments of the forces acting on the crane and suggest design changes to make it more stable.

An excellent answer will include a clear and detailed description of all the forces and their moments, and give the conditions necessary to make the crane topple together with improvements in the design to prevent this.

Examples of points made in the response:
- The load produces a turning effect/moment (clockwise).
- The counterweight on the opposite side to the load produces a turning effect/moment in the opposite direction/anticlockwise.
- The weight of the crane acts from the centre of mass.

- As the load moves horizontally its moment increases.
- When the clockwise and anticlockwise moments are not equal the crane will topple.
- When the moments are unbalanced there is a force that may bend the tower instead of toppling it.
- The crane will be more stable if:
  - the centre of mass is over the tower
  - the centre of mass is as low as possible
  - the base is as wide as possible
  - so that the line of action of the resultant force does not fall outside the base.

70. Electromagnets

1 (a) C (1)
  (b) the current in the wire (1)
2 The paper cone attached to the coil vibrates as the force attracting it to the permanent magnet varies (1) because the strength of the magnetic field depends on the current in the coil (1).
3 Advantages of electromagnet: force can be varied by changing current/you cannot make the permanent magnet drop the can (1). Disadvantages: electrical energy needed to provide power/permanent magnets do not need a power source (1). Conclusion: electromagnets are better as they can be controlled (1).

71. The motor effect

1 coil (1); 90° (1); up (1)
2 (a) Any two of: increase the number of turns of wire on the coil (1); increase the current in the coil (1); increase the strength of the magnet (1).
  (b) Reverse the direction of the current (1) or magnetic field (1).
  (c) The force on the top of the coil is upwards and the force on the bottom of the coil is downwards. The forces the coil are equal and opposite (1) because the current is moving in the opposite direction in the two parts of the coil (1).
3 When a current flows through the coil, the motor effect/a force makes it turn (1). The restoring spring stretches, producing an equal and opposite force (1) that stops the needle.

72. Electromagnetic induction

1 A: primary coil; B: secondary coil; C: iron core (all three correct, 2 marks; 1 correct, 1 mark)
2 (a) The needle moves in one direction when the magnet moves towards the coil (1) and in the opposite direction when it moves away from the coil (1).
  (b) When the magnet moves the coil 'cuts' the magnetic field/lines of force (1) and a potential difference/current is induced in the coil (1). *The answer must make clear that the potential difference is only induced when the magnet or coil are moving.*
3 The alternating potential difference produces a changing (1) magnetic field in the iron core (1), which induces a changing/alternating potential difference across the secondary coil (1).

73. Step-up and step-down transformers

1 more (1); greater than (1); step-down (1)
2 The number of turns on the secondary coil is five times bigger than the number of turns on the primary coil (1) so the output on the secondary coil is five times 10 V (1) = 50 V

3 (a) $I_p = V_s \times I_s/V_p = 230\,V \times 800\,A/33\,000\,V$ (1) = 5.7 A (1)
  (b) The transformer is 100% efficient (1).

74. Switch mode transformers

1 (5–6 ticks correct, 3 marks; 3–4 correct, 2 marks; 1–2 correct, 1 mark)

Statement	Traditional transformer	Switch mode transformer
works at mains frequency of 50 Hz	✓	
small and light		✓
converts mains electricity (230 V) to a lower potential difference	✓	✓
works at a frequency of 50 kHz to 200 kHz		✓
contains a heavy iron core	✓	

2 Traditional transformers use power when plugged in even if it is not being used (1), and waste energy or can overheat, causing fires (1). Switch mode transformers do not draw power when not being used/under load (1).
3 Any three points from: switch mode transformers do not use power when not in use (1) so do not overheat or waste energy (1); switch mode transformers do not have a heavy iron core (1) so are smaller/lighter (1).

75. Physics six mark question 3

A basic answer will give a brief explanation of how the circuit is broken.

A good answer will give some explanation of how the magnetic field in the coil operates the circuit breaker.

An excellent answer will give a clear and detailed explanation of the sequence of events that causes the circuit to break.

Example of points made in the response:
- Current flows through the coil.
- A magnetic field is produced.
- This attracts the soft iron bolt.
- As the current increases the strength of the magnetic field increases.
- But the bolt is held in position by its spring.
- When the current is overloaded/too high the force is stronger than the force from the spring,
- and the bolt is pulled toward the coil.
- The switch opens/the spring pushes the switch open.
- The circuit is broken/the current stops flowing.

Practice paper answers

Further Additional Science Biology B3 practice paper

1 (a) (i) A (1)
    (ii) B (1)
  (b) Active transport involves movement of particles against the concentration gradient/needs energy input/needs oxygen to be available (1).
2 (a) (i) vena cava (1)
    (ii) aorta (1)

(b) Any two from: pulmonary artery carries deoxygenated blood/pulmonary vein carries oxygenated blood (1); pulmonary artery has thick muscular wall/pulmonary vein has thinner wall (1); blood in pulmonary artery under higher pressure than blood in the pulmonary vein (1).

(c) The heart has valves (1); to prevent backflow of blood (1).

3 (a) transpiration (1)

(b) removal of anomalous result of 21 mm (1); mean value = (59 + 68 + 62) ÷ 3 = 63 mm (1); 63 mm in 5 minutes, so 63 ÷ 5 = 12.6 mm per minute (1)
If all four numbers are averaged, then 1 mark for 52.5 as the mean value, and 1 mark for 10.5 mm per minute.

(c) The distance moved would be shorter (1); less water evaporates in colder temperatures (1).

(d) The rate of photosynthesis goes down (1); so stomata in the leaves open less (1).

4 (a) small intestine (1)

(b) Any two pairs of marks from: thin (epithelial) surface layer of cells (1) means that rate of diffusion is high because the distance for diffusion is small (1); rich blood supply/large numbers of capillaries (1) maintains concentration gradient for diffusion by removing food molecules (1); large surface area (1) means that diffusion happens faster (1).

5 (a) (i) 29.0 °C ± 0.1 °C (1) *This means that 28.9 °C, 29.0 °C or 29.1 °C would all be correct.*

(ii) 80% female : 20% male (1); so, with 120 offspring, this is 96 females and 24 males (1)

(b) Increasing temperature reduces the proportion of males in the population (1) and could mean that there are no males, so the turtles could not reproduce (1).

(c) Carbon dioxide in the atmosphere is responsible for global warming (1); this gas can dissolve in sea water (1); as oceans are large, large amounts of carbon dioxide can be removed from atmosphere into the oceans (1).

6 (a) insulin (1); pancreas (1)

(b) glucagon released (1); converts stored glycogen into glucose (1)

(c) (i) need to reduce glucose concentration in the blood to minimum levels (1)

(ii) Any three from: starting level is within normal limits (1); in the test, the blood glucose concentration only rises to 8/does not rise above 9 (1); it then falls to normal levels (1); this can only be brought about by insulin release (1).

(iii) glucagon (1)

7 (a) Transport of carbon dioxide is in the plasma (1); in solution (1); transport of oxygen is in red blood cells (1); using haemoglobin (1).

(b) Platelets/cell fragments in the blood (1); come together in a wound to make the clot (1).

(c) Artificial blood can be given to any individual (1). The need for cross matching is eliminated (1).

8 (a) The diaphragm muscle contracts and flattens (1); the ribcage moves upwards and outwards (1); pressure reduces inside the thorax and air is drawn into the lungs (1); volume in the thorax increases (1).

(b) Any four from: blood vessels that supply capillaries in the skin dilate (1); so that more blood flows into the capillaries near the skin surface (1); and heat is lost from the body by radiation (1); sweat glands in the skin release more sweat (1); which also cools the body (as it evaporates from the skin) (1).

9 A basic answer would give a small number of facts, with little linking together or explanation of why these are advantages.

A good answer would give some explanation, although this is likely to concentrate on only one area, most likely the rate of the process.

An excellent answer would give a complete explanation of the advantages, considering both the speed and the energy factors involved.

Examples of points made in the response:

- Microorganisms grow very rapidly.
- Easy to find optimal growing conditions for microorganisms.
- Can be grown in many locations/climate not a factor.
- Can modify microorganisms to produce better product.
- Cheaper to use microorganisms.
- Mycoprotein contains enough of each main food group to fulfil requirements of a healthy diet.
- Low in saturated fat/cholesterol.
- Ethical concerns about eating meat.
- Farming animals inefficient.
- They need feeding on plant material.
- Energy transfer not efficient from food to animal.

Further Additional Science Chemistry C3 practice paper

1 (a) Both organised elements in atomic weight (mass) order (1).

(b) Mendeleev left gaps for elements yet to be discovered, OR he swapped some elements' positions (1), so that the elements appeared in the right positions (according to their properties) (1).

(c) in atomic number/proton number order (1)

2 (a) When they react with water they produce alkaline solutions (1).

(b) Atomic size increases down Group 1 (1) as more electron shells are occupied (1).

(c) Reactivity increases down Group 1 (1) as the higher the outer energy level the more easily an electron is lost, or as the outer electrons are further from the nucleus and so less strongly attracted (1).

3 (a) Ca^{2+} (1)

(b) Soaps form a scum in hard water (1) and don't produce a lather readily or they need a lot of soap to produce a lather (1).

(c) (i) They would add a small amount/1 cm^3 of soap solution at a time and shake (1). Note volume added when lather formed persists (1).

(ii) It tells us that water sample 1 is mainly temporary hard water (1) as the hardness is removed by boiling (1).

4 (a)

Energy in	Energy out
1 × C=C = 1 × 602 kJ = 602 kJ	4 × C=O = 4 × 798 kJ = 3192 kJ
4 × C—H = 4 × 414 kJ = 1656 kJ	4 × O—H = 4 × 458 kJ = 1832 kJ
3 × O=O = 3 × 497 kJ = 1491 kJ	
Total = 3749 kJ (1)	Total = 5024 kJ (1)

Energy change = 5024 kJ − 3749 kJ = 1275 kJ (1)

(b) The energy given out is greater than the energy put in (1), so overall energy is given out (1).

(c) $C_4H_8 + 6O_2 \rightarrow 4CO_2 + 4H_2O$ (1 mark for correct formulae and 1 mark for correct balancing)

5 (a) food flavouring or perfume (1)

(b) activation energy (energy needed to break bonds) (1)

(c)

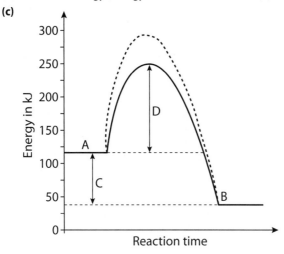

Dotted line starting and finishing at the same points (1) but with the peak above the one on the graph (1).

(d) The energy taken in (D) is less than the energy given out (D + C) (1); so overall energy given out and exothermic (1).

6 (a) (i) $2H_2(g) + O_2(g) \rightarrow 2H_2O(l)$ (1)

(ii) $C_2H_5OH(g) + 3O_2(g) \rightarrow 2CO_2(g) + 3H_2O(l)$ (1)

(b) The only product of burning hydrogen is water (1), which causes no pollution (1).

(c) 7 (1); hydrogen (1); alcohols (1)

(d) (i) ethanoic acid (1); $C_2H_4O_2$ (1)

(ii) oxidation (1)

7 (a) The reaction is reversible/an equilibrium reaction (1).

(b) If the gas pressure is increased it favours the formation of the least number of gas molecules (1); therefore there will be more ammonia left (1).

(c) Ammonia is soluble in water (1).

(d) (i) the Haber process (1)

(ii) methane and steam (1)

(iii) A lower temperature is not used as it would make the reaction too slow (1). A higher pressure would cost too much (1).

8 (a) The positive test for aluminium ions was shown by the addition of sodium hydroxide solution (1). The positive test for sulfate ions was shown by the addition of barium chloride/hydrochloric acid (1).

(b) (i) by adding an indicator (1) that will change colour when neutralisation occurs (1)

(ii) It is a rough/anomalous value (1).

(iii) volume = 9.9 cm³ ÷ 1000 = 0.0099 dm³; number of moles of HCl = $N = C \times V = 0.1 \times 0.0099 = 0.00099$ (1). 1 mole of HCl neutralises 1 mole of NaOH ∴ 0.00099 moles of NaOH; ∴ 0.00099 = $C \times 0.02$ ∴ $C = 0.00099/0.02 = 0.045$ mol/dm³ (1).

9 A basic answer would give a simple description of how flame tests and the addition of sodium hydroxide solution can be used to identify certain metal ions.

A good answer would give a clear description of how flame tests and the precipitates formed by sodium hydroxide solution can be used to identify certain metal ions, including examples of significant positive observations.

An excellent answer would give a detailed description of how flame tests and the precipitates formed by sodium hydroxide solution can be used to identify certain metal ions. It also includes significant positive test results, an explanation of chemical reactions involved and some evaluation of the reliability of the tests.

Examples of points made in the response:

- Flame tests are carried out by dipping a rod in unknown solution and holding it in a blue Bunsen burner flame.
- The colour of the flame can be used to identify certain metal ions, e.g. yellow flame = sodium and red flame = calcium.
- Not totally reliable as certain colours are produced by more than one metal.
- Adding sodium hydroxide solution to certain metal ion solutions produces a precipitate, e.g. $NaOH(aq) + CuSO_4(aq) \rightarrow Cu(OH)_2(s) + Na_2SO_4(aq)$. (The precipitate is the metal hydroxide.)
- Metals can be identified by the colour of the precipitate, e.g. blue precipitate = copper(II) ions and white precipitate = aluminium, calcium or magnesium ions.
- Not totally reliable as certain colours of precipitate are produced by more than one metal.
- There are also problems with both tests if there is a mixture of metal ions present.

Further Additional Science Physics P3 practice paper

1

Statement	X-rays	ultrasound	
They are a form of electromagnetic radiation.	✓		(1)
They are a form of wave motion.	✓	✓	(1)
They can cause ionisation in cells.	✓		(1)
They are reflected at the boundary between two different tissues.		✓	(1)
They travel at the speed of light.	✓		(1)

2 (a) It improves the accuracy of the measurement (1).

(b) rearranging gives $f = 1/T$ (1) = 1/2.4 (s) (1) = 0.42 (1) Hz

(c) (i) A, C and E (1)

(ii) The period of the pendulum increases as the length of the string increases (1). *Saying the period is proportional to the length is incorrect.*

(d) Changing the mass has no effect on the period (1).

3 (a) investigating the interior (1) of bodies or other objects

(b) The light undergoes total internal reflection (1) because it hits the boundary at an angle greater than the critical angle (1). *The marks may be awarded for a labelled diagram showing this.*

(c) refractive index = 1/sin c = 1/sin 40° (1) = 1/0.643 = 1.56 (1)

4 (a) When light enters the lens it is refracted towards the normal (1). When it leaves it is refracted away from the normal (1).

(b) power = $1/f$ = 1/0.15 (1) = 6.67 or 6.7 (1) dioptres

(c) Diagram completed as shown overleaf. 2 or 3 rays drawn correctly (1) from point on object meeting at point on image (1). Image 3 times height of object and 3 times distance from lens (1). *Rays projected back behind the lens should be shown as dashed lines but marks are not deducted if this is not done.*

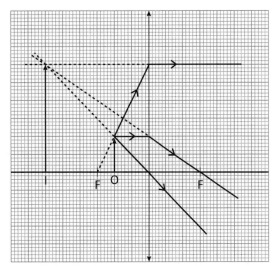

(d) magnification = height of image/height of object
3 cm/1 cm (1) = 3 (1)

5 (a) The force on the piston increases the pressure (1) in the liquid, which is transmitted in all directions/to the liquid in the needle (1).

(b) pressure on piston = 0.5 N/(4 × 10⁻⁴) m²
= 1.25 × 10³ Pa (1); force on cap = pressure × area
= 1.25 × 10⁻³ Pa × 1 × 10⁻⁶ m² (1) = 1.25 × 10⁻³ or 0.00125 (1) N

6 (a) The centre of mass is the point at which the mass of the object seems to be concentrated (1).

(b) make the base wider (1); make the base thicker so the centre of mass is lower (1)

(c) The flowers make the centre of mass higher (1) so when it is tilted the line of action of the weight (1) is more likely to fall outside the base (1).

7 (a) (i) downward arrow (1) (on the right-hand side of the coil)

(ii) when the coil is vertical (1)

(b) (i) Any two from: stronger magnetic field (1); larger current (1); more turns on coil (1).

(ii) reverse the direction of the current/magnetic field (1)

8 (a) (i) There are fewer turns of wire on the secondary than on the primary coils (1) (or there are more on the primary).

(ii) rearranging gives $V_s = V_p \times n_s/n_p$
= 230 V × 24/1120 (1) = 4.9 (1) V

(b) power output = $V \times I$ = 24 V × 0.1 A = 2.4 W (1); current from mains = 2.4 W/230 V = 0.010 or 0.0104 (1) A

(c) The new charger is lighter (1), smaller (1) and does not use power when not charging the batteries (1). *That switch mode transformers work at high frequency is true but not an answer to the question.*

9 $M = F \times d$; anticlockwise moment = 5 N × 1.2 m = 6 N (1) Nm; clockwise moment = W N × 0.2 N (1). When the plank is balanced clockwise moments = anticlockwise moments; 6 Nm = W N × 0.2 m (1); W = 6/0.2 N = 30 N (1).

10 A basic answer will describe the function of one part of the eye and camera, and state that an image is formed in both.

A good answer will describe the functions of most of the parts of the eye and the camera, and suggest how the image is formed in both.

An excellent answer will compare and contrast the functions of the parts of the eye and camera, and give a detailed description of the formation and properties of the image formed.

Examples of points made in the answer:
Function of parts of the eye:
- The cornea refracts/focuses light as it enters the eye.
- The iris controls the size of the pupil,
- controlling the amount of light entering the eye.
- The lens refracts/focuses the light to form an image.
- The shape of the lens is controlled by the ciliary muscles and suspensory ligaments.
- The image is formed on the retina which sends nerve signals to the brain.

Function of parts of the camera:
- Aperture controls the amount of light entering the camera,
- like the iris/pupil.
- The lens bends/focuses the light to form an image,
- like the cornea and lens.
- The image forms on the film/CCD,
- which is equivalent to the retina recording/transmitting the image to the brain

The image in both is:
- real
- inverted
- smaller/diminished.

General points:
- The lens in both eye and camera can be adjusted to produce a sharp image of objects at different object distances.
- The lens does not change shape in the camera but does change distance from the film/CCD.

This page has been left deliberately blank.

This page has been left deliberately blank.

This page has been left deliberately blank.

Published by Pearson Education Limited, Edinburgh Gate, Harlow, Essex, CM20 2JE.

www.pearsonschoolsandfecolleges.co.uk

Copies of official specifications for all AQA qualifications may be found on the AQA website: www.aqa.org.uk

Text and original illustrations © Pearson Education Limited 2013
Edited by Jim Newall and Florence Production Ltd
Typeset and illustrated by Tech-Set Ltd, Gateshead
Cover illustration by Miriam Sturdee

The rights of Iain Brand and Peter Ellis to be identified as authors of this work have been asserted by them in accordance with the Copyright, Designs and Patents Act 1988.

First published 2013

17 16 15 14 13
10 9 8 7 6 5 4 3 2

British Library Cataloguing in Publication Data
A catalogue record for this book is available from the British Library

ISBN 978 1 447 94216 0

Printed in UK by Ashford Colour Press Ltd., Gosport, Hampshire.

All images © Pearson Education

Every effort has been made to contact copyright holders of material reproduced in this book. Any omissions will be rectified in subsequent printings if notice is given to the publishers.

In the writing of this book, no AQA examiners authored sections relevant to examination papers for which they have responsibility.

Contents Page

Published by Step Forward Publishing Limited
St Jude's Church, Dulwich Road, Herne Hill, London, SE24 0PB
Tel: 020 7738 5454 www.practicalpreschool.com
© Step Forward Publishing Limited 2008   Illustrations by Cathy Hughes

ISBN 13: 978 1904575 47 4

Stepping Stones to Creativity

Stepping Stones to Creativity is a series of books that aim to provide early years practitioners with a treasure trove of practical activities and resources to help develop the budding creativity of children in their care. Each book focuses on a different area of creativity, which is explored through forty of the most popular early years topics. In each topic you will find activities that support the five Early Learning Goals of Creative Development, so that you can ensure you are meeting the requirements of the Early Years Foundation Stage.

Many of these activities can be adapted for even younger children or extended to benefit Key Stage One children.

Creativity In The Classroom

Creativity is an elusive term. Although great enjoyment can be taken from working with imaginative children in the Early Years who are naturally eager to explore, in practice there are some key issues which face the practitioner:

What is 'creativity'?

In a nutshell, it is the ability to use knowledge and skills, plus a healthy dollop of imagination, to tackle and solve any problem. It is about taking risks and being involved in the learning process. Creative thinking does not just apply to obviously 'creative' tasks such as art or music, but can be used in investigations in science and across the foundation stage curriculum. To develop creativity children need time, space, and multiple opportunities to experiment with materials and ideas. They also need to be encouraged to make connections between ideas as they play.

How can we as practitioners make this happen for the children in our care?

Young children are curious by nature. They learn by exploring and experimenting and 'having a go'. As practitioners we need to provide a stimulating environment together with lots of opportunities and of course unlimited time! Although it can be a challenge in multi-purpose buildings, a creative environment can be created by leaving art materials and musical instruments out for children to play and experiment with.

Does there always need to be an 'end product' to creativity?

Practitioners and parents often both fall into the trap of only valuing an 'end product' as proof of creativity. How many of us have heard an anxious parent berating their young child with the words, 'have you made me a painting this morning Tom?' Children sometimes learn to rush to the painting easel and apply a few hurried strokes of paint to appease their carer, even though they may have been involved all morning playing imaginatively and creatively in the role-play area. We need to appreciate creative play and processes as just as important as any finished artwork!

What is the value of creative group work?

In many settings, children often work together to produce joint artwork and models, as well as singing, drama, dance and music performances. This can be seemingly problematic, as at times the individual creative process may have to be subordinated to the purpose and will of the group. However, group creative work can provide an ideal opportunity to develop children's social and co-operative skills. More often than not it requires more imagination and creativity to work as a group!

What can practitioners do to develop children's creativity?

You may now be feeling how do I encourage children in my care to develop their creativity? The following are useful hints for nurturing young children's creativity:

- Provide sufficient time and opportunities for children to explore, experiment and practice their skills. Allow children time to work at their own speed. Try to avoid them being pressured by other children eager to try another activity.
- Some children require more encouragement to 'have a go' on their own. They may need you to suggest ideas, stimulate their imagination, and encourage them. Children will discover a lot through their own explorations, but unless there is an adult on hand to talk about their discoveries, learning opportunities can be missed.
- Children need to feel secure. They need to know that help is available if and when they need it. The tricky job is judging when to intervene if a child is struggling. Sometimes the process of problem-solving is part of the learning process, but do be prepared to model and teach new skills that children may require in order to progress. Experience will help you maintain the balance between being intrusive, and avoiding the frustration children feel when their own efforts are thwarted.
- Supply good quality resources – both materials and people! Challenge children by inviting artists or performers to show children their own particular area of creativity. In particular make use of talented parents who are willing to help.
- Encourage children to talk to each other about their work. Ask them to share how they overcame problems when constructing a model robot, or why they decided to make a loud sound at the end of the music.
- Try to be creative yourself! How long is it since you found time to develop your own creative gifts? Challenge yourself to learn a new skill this year.

Exploring Drama and Role-Play

Creative development is one of the six areas of learning in the Early Years Foundation Stage – the curriculum for all children under the age of five. The Statutory Framework, published in 2007, breaks down Creative Development into the four following aspects.

The Four Aspects of Creative Development:

Being Creative – Responding to Experiences, Expressing and Communicating Ideas

Corresponding E.L.G.: Respond in a variety of ways to what they see, hear, smell, touch or feel.

Corresponding E.L.G.: Express and communicate their ideas, thoughts and feelings by using a widening range of materials, suitable tools, imaginative and role-play, movement, designing and making, and a variety of songs and musical instruments.

Exploring Media and Materials

Corresponding E.L.G.: Explore colour, texture, shape, form and space in two and three dimensions.

Creating Music and Dance

Corresponding E.L.G.: Recognise and explore how sounds can be changed, sing simple songs from memory, recognise repeated sounds and sound patterns and match movements to music.

Developing Imagination and Imaginative Play

Corresponding E.L.G.: Use their imagination in art and design, music, dance, imaginative and role-play and stories.

Interestingly, the actual word 'drama' is not included in the Early Learning Goals for Creative Development but it clearly underlies a great deal of the goals' intentions.

A child is surely 'being creative' when, through drama and role-play, they are able to show a personal response to a dramatic, if pretend, situation. Drama games and skills enable children to 'express and communicate' their own ideas in a fun, imaginative and creative way. In an increasingly technological world children are spoonfed acceptable responses through exposure to passive entertainment in the form of television programmes and computer games. The world of drama and role-play can provide a vital medium for the development of imagination and imaginative play.

Drama and Role-Play

All practitioners desire to extend children's creativity by supporting their natural curiosity through play and exploration. Role-play areas should be inspirational, open-ended environments that enable children's creative learning, encouraging them to feel safe and secure as they extend their experiences of life. These environments offer many opportunities to develop cross-curricular learning, in particular language development, awareness of a variety of cultures, and knowledge and understanding of the world around us.

Using the activities in this book

Drama games
These games are great as starting points for a drama session, or to introduce a new topic to your children in a creative way. Many of them are fun 'warm-up' games that help the children to relax and feel comfortable with each other so that they are confident and able to express themselves in a non-threatening environment. These may be thinking word games, informal circle games, or energetic physical activities, which require more space and warm up bodies as well as minds.

Mime
The skills of mime require a great deal of concentration and these focused activities will develop children's ability to use their imagination and their bodies to tell stories and describe thoughts and feelings. Many of the games and activities can be adapted for use in other topics, for instance the game Mirrors works well across the topics of 'Food', 'Clothes' and 'Ourselves'.

Drama skills
This is the most varied section, which includes the development of some specific dramatic skills such as 'freeze frames', 'hot-seating', retelling stories, changing the endings of stories, and character studies. This section is called 'Drama skills' as the activities involved often extend a skill learned earlier in a drama game or mime activity. There are lots of suggestions for improvised dramas, which take the children beyond pretend play into stories with characters, dilemmas and crucially, resolutions. These are most effective if the teacher is prepared to go 'in role' and join in the drama to sensitively direct where necessary.

Role-play
Each topic is also matched to a suitable role-play scenario that can be set up simply in most settings. Instructions are included for the equipment required, as well as ideas for making homemade features. Possible roles that the children might like to explore are listed, along with lots of ideas for stories to be acted out, with or without teacher input.

Animals

Drama games

- Sit in a circle and pass round animal sounds: hissing, roaring, squeaking, mooing, and so on. Try not to repeat a previous sound.
- *Noah's Ark:* Make a collection of cards showing matching pairs of animals and hand them out to children. Ask the children to move around the room making the sounds and actions for their animal and try to find their partner to go into the ark.

Mimes

- *Animal antics:* In the style of the game 'beans' invite children to mime different animals. Call out 'cheetah' and ask the children to run fast around the room on all fours. Try 'snake' – sliding along floor; 'rabbit' – bunny hops; 'elephant' – move slowly, swinging a trunk; 'parrot' – fly around the room; 'crocodile' – snapping jaws, and 'horse' – children find a partner and trot around like a pantomime horse! Can the children think up some more animal mimes of their own?

Drama skills

- Use animal puppets to help children make up stories about animals from traditional tales or picture books.
- *Visit to the zoo:* Improvise a family trip to the zoo. Introduce the characters in the family. Talk about the preparation for the trip and the journey. At the zoo set up some situations or problems for the children to face in the drama such as: an escaped lion, the zookeeper needs help feeding the penguins, or one of the animals is sick.

Role-play

The vet's surgery

Set up: A waiting room with reception area, telephone, diary, computer-screen, posters, leaflets, pet food, chairs, a consulting room with table, scales, medical kit, soft toy animals and carrying boxes or baskets.

Roles: Vet, nurse, pet owners, receptionist, animals.

Stories: Cat escapes in the waiting room; poorly new pet; animal won't keep still to be examined; animal is too big to fit in the room (e.g. a horse or kangaroo); vet is scared of snakes! Read and act out the story *Mog and the V.E.T.* by Judith Kerr.

Autumn

Drama games

- *Firework display:* Sit in a circle and pass firework sounds around the circle. Try not to repeat any words or sounds. Then add body actions and sounds and perform a firework display.
- *Vegetable soup:* In the style of 'fruit salad' stand in a circle and give each child the name of a vegetable such as carrots, parsnips and pumpkins. Call out a vegetable and those children must change places in the circle. When you call out 'vegetable soup' everyone has to change places!

Mimes

- Mime making soup. Go through the process stage by stage – washing and chopping vegetables, cooking, blending, eating.
- Divide the children into two groups. Ask one group to make the shapes of trees, gently swaying in the wind. As the wind gets stronger the leaves begin to fall (fingers fluttering down to the ground) and acorns and conkers drop off the trees (fists bang on the floor). The other group are the quick squirrels and spiky hedgehogs who are searching for food.

Drama skills

- *The bonfire party:* Improvise a drama at a bonfire party. Talk about the characters who might be at the party. Plan and prepare the food. Build the bonfire. Discuss fireworks and the firework code. Carefully pretend to set off the fireworks. Use some of the children's firework

sounds and actions from the firework display game. Carefully introduce some possible problems to the drama such as a scared pet or an injury.

Role-play

Autumn glade

Set up: Simple home corner surrounded by tall trees made from corrugated cardboard trunks. Lots of real autumn leaves strewn about on top of a mat with a scattering of conkers, acorns, and other autumn treasures. Soft toy squirrels, owls, rabbits, mice, hedgehogs etc.

Roles: A family to live in the house, woodland animals etc.

Stories: An Autumn walk; hiding and finding Autumn treasures or food; listening to the sound of scrunching leaves; making a wish when catching a falling leaf; finding and caring for an injured animal. Act out the story of *Pumpkin Soup* by Helen Cooper.

Bears

Drama games

- *Share-a-bear:* Sit in a circle and pass a special 'shared bear' around. Encourage children to feel that when holding the bear they can share some happy or sad news with the group.
- *Bear Hugs:* Ask children to bring in bears from home. Move around the room with their bears as you shake a tambourine. When you tap it they must 'hug' the nearest bear and its owner, very tight!

Mime

- *Sleepy Bear's Honey Pot:* Sit in a circle and invite one child to be Sleepy Bear in the middle with a small pot placed behind him or her. Point at another child to creep into the circle and take the pot without waking Sleepy Bear. Make it more difficult by putting something in the pot that makes a sound when moved!
- *Bears:* Play this mime game in the style of 'beans'. Start by introducing two or three different bears such as 'Black Bear' – rear up on hind legs and growl, 'Brown Bear' – curl up and hibernate and 'Polar Bear' – slide around on all fours. Later add 'Spectacled Bear' – make glasses using hands, and 'Teddy Bear' – hug the nearest person.

Drama skills

- *We're All Going On A Bear Hunt:* Act out the well-known rhyme by Michael Rosen.
- Talk about the three bears from the story of Goldilocks. Practise using voices in different ways to represent the three bears: Daddy Bear's low growly voice, Mummy Bear's middle normal voice, Baby Bear's high squeaky voice. Play who's coming to breakfast? Ask the question together and take turns to reply using one of the bear's voices. Can the children guess which bear is coming to breakfast?
- Help children to act out the scene from Goldilocks when the three bears return to the house to find there has been an intruder! Interview characters from the story including Goldilocks. Make up a new ending and act it out.

Role-play

Three Bear's House

Set up: Three different sized bowls, chairs and beds, bear masks or dressing-up suits, other home furniture.

Roles: Goldilocks, three bears, other traditional story characters who could visit the house such as Jack, the big bad wolf, Hansel and Gretel, Little Red Riding Hood.

Stories: Mix and match Goldilocks with other fairy tales.

Clothes

Drama games

- *Pass the hat:* Sing this song as you pass a hat around the circle:

  Who will wear the hat today?   X3
  Who will wear the hat?
  [Tune: In and Out the Dusky Bluebells]

  Whoever is holding the hat at the end of the song has to put it on and assume the character and add a line of dialogue if possible. Try it with two different hats and make up a conversation.

Mime

- *Costume box:* Pack a small suitcase with items of clothing and dressing up clothes. Let children take turns to choose an item to wear and then mime a character or activity for the others to guess.
- *Mirrors:* In pairs ask children to take turns miming putting on different clothes for their partner to mirror. Try hat and scarf, gloves, coat and shoes. See also the Food and Ourselves sections.

Drama skills

- Go on a journey and change clothes according to the different places you visit. Sort out a pair of boots and an umbrella, a sun hat and glasses, hat and scarf, and see how quickly the children can change as they visit the rainforest, the desert and the north pole.
- *Magic hat, slipper or cloak:* Present the clothing to the children and explain that it has magic qualities in that it can make you invisible, clever, able to fly and so on. In small groups improvise drama situations using the magic clothing.

Role-play

The clothes shop

Set up: Clothes on hangers sorted into types and sizes, cash tills, purses, accessories, money, carrier bags, a changing room, a full-length mirror. Change the setting to a shoe shop and add lots of different types of shoes in boxes and shoe-measuring equipment.

Roles: Shop assistant, customers.

Stories: A rude customer or shop assistant; a hard to please customer; a customer who doesn't fit any of the clothes; returning a faulty item; a fashion show. Act out the traditional story of *The Elves and The Shoemaker*.

Colours

Drama games

- Develop speaking and listening skills by making up colourful tongue-twisters such as 'big beautiful blue balloon burst with a bang' and 'red pepper, yellow pepper, green pepper, blue, pick a coloured capsicum and cook it for me too'.
- *Changing places:* Stand in a circle and call out "change places if you're wearing red". Choose different colours and see how quickly the children can move.
- *Introductions:* Go round the circle taking turns to introduce yourself and tell the group what your favourite colour is. For example 'my name is Joel and my favourite colour is purple.'

Mime

- *Moody colours:* Talk about colours and moods. Choose facial expressions for each colour or mood: red for angry, blue for sad, green for jealous, yellow for scared, orange for happy, and black for moody. Show a coloured card and ask the children to respond. Extend the activity by thinking up actions for each colour as well, such as jumping for orange and stamping for red.

Drama skills

- Act out the story of *Little Red Riding Hood*. Would the story change if the colour of the cloak was different?
- *Colourful dramas:* Extend some of the ideas from Moody colours into paired dramas. Each child takes on a colour to reflect their mood and make up a story. What might happen when the blue boy meets the yellow boy? Could the red girl be cheered up by the orange girl?

Role-play

Colour swap shop

Set up: Toys and different items sorted into colours, coloured cards, shelving, signs in different colours.

Roles: Shopkeeper, customers, Rainbow Fairy.

Stories: In the swap shop children can swap coloured cards for an item of the same colour. The naughty Rainbow Fairy comes along and mixes all the colours up!

Dinosaurs

Drama games

- *Action/Freeze:* Ask children to find a space in the room and when you shout 'action' to move freely around taking care not to bump into anybody. When you shout 'freeze' they must all stand still like a statue. Then ask them to work in pairs and create contrasting dinosaur statues – tall and short, big and small, carnivores and herbivores, spiky and smooth, fast and slow.

Mime

- *Moving dinosaurs:* Talk about different sorts of dinosaurs such as carnivores, herbivores, flying dinosaurs and so on. Ask children to think of different ways to move for each dinosaur.

Drama skills

- *The dinosaur egg:* Improvise a drama about a group of children who discover a large plastic egg while on a picnic. Talk about the characters and plan and prepare the picnic food. Then travel to the picnic in the forest once you find the egg let the children decide what to do with it. Should they leave it alone, take it home, or break it? Narrate the ending – make sure the egg is returned to where they found it so that when the mother dinosaur comes back to find it out hatches a baby dinosaur!
- *We're all going on a dinosaur hunt:* Adapt the well-known rhyme about the bear hunt to include dinosaurs!

Role-play

The dinosaur museum

Set up: Create displays of dinosaur bones or pictures. Tickets, cloakroom, leaflets, a recorded museum guide, museum shop with toys and postcards. Make model dinosaurs as exhibits from wire, modroc, papier mache and paint.

Roles: Curator, tour guides, visitors, teacher, school party, dinosaurs.

Stories: Night at the museum – the dinosaur exhibits come to life! A school trip to the museum in which a child gets lost or breaks an exhibit. Set up a new display with an exciting new dinosaur that has just been discovered.

Families

Drama games

- *Family voices:* Encourage children to use different voices to answer the question 'can you use a daddy voice?' Change the question to mummy, baby, granddad, angry aunty, unhappy uncle, silly sister, bossy brother etc.
- *One-line family characters:* Sit in a circle and ask children to think of something mum or dad might say. Go round and invite them to deliver a 'one-line character' for others to guess. Extend by adding a mood such as angry, sad, happy or impatient.

Mime

- *Family freeze frames:* Explain that you are going to arrange the children into family groups and then take pretend pictures of them. Ask each group to talk and interact with each other and then shout 'freeze' as a signal to be still. Try inventing different family situations such as a celebratory meal, an outing, an argument etc.

Drama skills

- *Two Fat Gentlemen:* Learn the finger rhyme, as shown below, and then use as a framework for different family characters such as: two old granddads, two thin mummies, two tall daddies, two teen sisters, and two naughty babies. Number the characters 1 – 5 and then mix and match the numbers to create new dramatic situations where number 1, the old granddad, meets number 4, the teen sister, and so on.

The Two Fat Gentlemen Finger Rhyme

Two fat gentlemen
Met in a lane

Bowed most politely
And bowed once again

And said 'How do you do?'

And 'How do you do?' again.

 Stepping Stones to Creativity

Role-play

The home corner

Set up: Usual home corner furniture, baby dolls, camera, presents, baby equipment: food, clothes, wipes, toys, bottles, high chair, buggy, cot, changing-mat, bath, towels.

Roles: Mum, dad, children, baby, grandparents, aunts, uncles, neighbours and friends.

Stories: New baby in the family. Act out the baby coming home: how does everyone in the family feel? The baby cries a lot and nobody can sleep! Everyone has to be quiet so baby can sleep and every little noise wakes him up! Tidy up the house ready for visitors. Organise a celebration or party in the house.

Farms

Drama games

- Sit in a circle and pass round farm animal sounds trying not to repeat each other. Add suitable actions.
- *Action/freeze:* Ask children to find a space to stand in. Make sure they cannot touch anybody near them and that they have plenty of space to move around. When you shout out 'action' invite them to move around freely until you shout 'freeze'. Then they must freeze and pretend to be a farm animal statue or scarecrow of their choice.

Mime

- *Down on the farm:* Ask children to take turns miming different farm animals for the group to guess. Add a farmer and mime how we use the different animals on the farm such as milking a cow or riding a horse.
- Pass a pretend piece of food around the circle. Start with an orange that needs to be peeled and eaten segment by segment. Pass it to the next child and let them mime a different piece of food such as a crunchy carrot, banana, or strawberry. Can the others guess what is being eaten each time?

Drama skills

- *Scary scarecrows:* Talk about scarecrows. Ask the children to pose as a scarecrow; very still but with a scary face! Improvise a dramatic story about scarecrows who come to life. Are they friendly or not? Go into role as a chief scarecrow and direct the story if necessary. The

scarecrows could work together to save the farm from being built on, or gang together to scare away the farmer.

Role-play

The farm and farm shop

Set up: Home corner with lots of cooking utensils, a model fire made from tissue paper, cellophane, and newspaper. An outside area with sit-and-ride toys including a tractor, a field of crops made from corrugated cardboard. A farm shop with shelves, tables, real or pretend fruit and vegetables, bread, homemade biscuits, plant pots, tools etc. Make fruit using screwed up newspaper, modroc and paint. Make real biscuits to sell at the farm shop.

Roles: Farmer, farmer's wife, children, workers, animals, shop assistants, customers, school children.

Stories: School trip to visit the farm or shop; a lost animal; the discovery of a new mystery fruit or vegetable; harvesting crops on the farm for the harvest festival; making bread. Act out the traditional stories *The Little Red Hen* or *The Enormous Turnip*.

Flight

Drama games

- Ask children to stand in a space in the room. Explain that they are to move around slowly and carefully, not bumping into anybody else when they hear the slow drum beat. As the tempo changes and the drum beat speeds up ask them to pretend to fly around on tiptoes still taking care not to touch anyone else.
- *Balloons:* Blow up a balloon as children watch and then release the air carefully. Ask the children to pretend to inflate as you blow into the balloon and then slowly deflate. Repeat, and this time let the balloon go so that it whizzes around! Can they do the same and add sound effects?

Mime

- *Magic carpet:* Sit in a circle and ask the children to close their eyes as you narrate a journey on a magic carpet. Start with a flight to the children's own bedrooms to select a favourite toy or book. Describe where the carpet is flying over and end the journey with a recognised phrase such as 'the carpet landed with a bump and a tumble!' Go on flights to other places and countries. This is a really good way of helping children to concentrate and enter their dramatic imagination.

Drama skills

- *In a hot air balloon:* Improvise a drama from the air. A group of characters go on an early morning flight in a hot air balloon and witness something bad happening on the ground below. Can they do anything to stop it?

Role-play

The airport

Set up: An airplane – put pairs of seats in a row with an aisle down the middle. Open a check-in desk with tickets, passports, uniforms, a luggage conveyer belt, an x-ray machine for hand baggage. Open a shop or café in the airport waiting area.

Roles: Pilot, stewards, passengers, airport workers, porters.

Stories: Late to catch a plane, delays to flight and have to sleep in the airport, grumpy passengers, find something strange in a passenger's luggage.

Food

Drama games

- *I went to the market:* This is a good game for concentration and listening skills. Go round the circle asking each person to add an item of food they bought at the market to the list. See the topic on Shopping for related games.
- *Fruit salad:* Stand in a circle and give each child the name of one of three or four different fruits such as apples, grapes, oranges, bananas. When you call out each type of fruit they must change places. On the signal 'fruit salad' everybody must move!
- *Beans:* Use the names of different beans to initiate actions: runner beans – run on the spot; broad beans – stand with hands and feet spread side; string beans – stand with body stretched as tall as possible; jelly beans – wobble like jelly; baked beans – curl up small on the floor, and so on.

Mime

- *Mirrors:* Work in pairs and mime eating food in a mirror. Try messy food such as pasta, packets of crisps or chewy toffee. Partners must try to mirror all the facial expressions and hand movements. See the topics of Clothes and Ourselves.
- *Chewing gum:* Pretend to chew gum with your mouth closed, stretching all the facial muscles and pulling lots of silly faces. Ask children to join in and then remove gum and pretend that it's stuck to your finger and won't shake off. Have fun as the gum gets stuck to your toes, nose, ear, legs and use as a general physical warm up!

Drama skills

- *The market stall:* Improvise a drama about a busy market stall selling lots of different food. Help children to make up market calls for their wares. Encourage them to project their voices. Introduce situations involving difficult customers, a thief and bad weather.

Role-play

Café or restaurant

Set up: Tables, chairs, tablecloths, menus, vases of flowers, real and pretend food, crockery, and a blackboard for specials! Organise a kitchen area with a cooker, sink, place to prepare

- *Friendship chair:* Take turns for two children to sit on a chair in the middle of the circle. Ask them to mime something strange happening to the chair e.g. the chair is burning hot, shrinking, has itching powder or a sharp pin on it, a broken leg, it's too hard, too soft, too small, grows wings and flies away, and so on.

Drama skills

- Use a story, poem or real situation to compose a dramatic argument between two friends. Talk about what can cause arguments. Help children to work in pairs and create an argument. How will they resolve the problem? Remind them that it is only 'pretend'.

Role-play

A playground or park

Set up: Climbing and play equipment, a balance-beam, plastic hoops, balls, quoits, sand pit, bench or seating, an ice-cream van or stall, a pond, sit-and-ride toys, park keeper's hut.

Roles: Children, families, a park keeper.

Stories: Friends are playing at the park: one gets hurt or injured on the equipment; one loses their belonging; they lose each other or an adult; they feed the ducks; find something strange in the park; encounter a grumpy park-keeper. Act out stories from the *Percy the Park Keeper* series by Nick Butterworth.

food etc. Introduce a particular cuisine such as Chinese, Indian, Italian or healthy food. Change into a fast-food restaurant.

Roles: Waiter, waitress, chef, customers.

Stories: The kitchen runs out of food, the chef throws a tantrum when criticized, a rude or lazy waiter, a difficult customer, a customer has forgotten their purse and can't pay, a restaurant is threatened with closure.

Friends

Drama games

- *Meet and greet:* Ask children to find a space and to then move around, taking care not to bump into anybody. When they hear you call 'meet' they must meet the nearest person and 'greet' them in different ways. Try shaking hands, bowing, hugging, high-fiving, and so on. Vary the type of greeting each time.

Mime

- *Throw that feeling:* Sit in a circle and throw a beanbag to one child. Ask them to then mime a 'feeling' such as sad, happy, scared, angry, excited, worried, calm, tired and so on, just using their face. Then ask them to throw the beanbag on to a friend and call out a different feeling.

Gardens

Drama games

- *I went to the garden:* Use the well-known rhyme to act out planting and watering plants, and getting rid of pests, as you grow plants in the garden.
- *Lists:* This is a word game that can be adapted to lots of different topics. Go round the circle naming different flowers and plants from the garden. Take care not to repeat one already named.

Mime

- Choose three or four different activities that might happen in a garden such as: playing in a paddling pool, weeding the flower bed, mowing the lawn, playing on the swing, playing hide and seek. Ask children to mime each of them and change quickly between them when you call out the instructions.

Drama skills

- Improvise a drama about someone who needs help in their garden. Go into role as an old person and describe your garden: the long grass, lots of weeds, a broken fence, huge stones in the soil etc. Organise the children into three or four groups and get them to help with each of the problems. How will they get rid of the stones? Will they build a wall or rockery, move them in a wheelbarrow or bury them. Work with the children's ideas. Add tension by saying the work has to be completed before it rains.
- Act out the story of *Jasper's Beanstalk* by Nick Butterworth as Jasper tries to grow a plant and instead grows more and more impatient. Compare this with the traditional story of Jack and the Beanstalk.

Role-play

The garden centre

Set up: Real and pretend plants, tools, flowers, shopping baskets, plant pots, watering cans, seed packets, garden furniture, sand, gravel, buckets, wheelbarrows, aprons, gardening gloves, boots. Make your own seed packets using cut-out pictures from seed catalogues and sealed envelopes full of rice.

Roles: Manager, shop keepers, customers, gardeners.

Stories: A very busy shop with not enough staff; a naughty dog loose in the store; flowers and plants start dying mysteriously; a wheel falls off a wheelbarrow, and so on.

Growth

Drama games

- *Breathing:* Ask children to stand up straight and practice breathing in and out. Ask them to breathe in to the count of 3, hold their breath for 3 and then breathe out. Make sure they don't lift their shoulders. When they are confident with this ask them to breathe out through different lip shapes such as 'oo', 'ah' and 'ee'. Lengthen the time that children hold their breath to the count of 5 and then 10.

Mime

- Try some physical drama about a growing seed. Ask children to curl up small like a seed. Very slowly introduce change as they begin to grow shoots and grow up into a plant with flowers, or a tall tree. Make all the movements tiny and very slow, as that is more challenging. Use slow, changing music to help with the concentration.
- *What's my line going to be?:* Talk about what jobs the children would like to do when they grow up. Ask them to mime the job for the others to guess. See also the topics of Ourselves and Night.

primroses

Drama skills

- *Jack and the Beanstalk:* Retell the traditional story. Talk about the different characters : Jack, Jack's mum, the cow, the bean seller, the giant and the giant's wife. Divide the story into scenes and act out the story. Interview characters from the story. How did Jack's mum feel when he brought home a handful of beans? How did the giant's wife feel when Jack came back the second time? Was the giant really bad or was he just protecting his property? Who was the mysterious bean seller?

Role-play
The giant's castle

Set up: Oversized furniture and adult clothes. A bag of coins, a soft-toy hen, a miniature harp made from cardboard and strings. Lots of places to hide.

Roles: Jack, giant, giant's wife.

Stories: Jack and the beanstalk, hide and seek, the adventures of the giant before Jack came to visit. Where did the giant get the magic hen from?

Holes

Drama games

- *The answer is..:* This is a great game for trying to keep a straight face, which children really enjoy. Choose a word such as 'doughnuts' and agree that whatever question you ask the answer must be 'doughnuts'. Then ask questions such as 'what do you wear to go to bed?' and 'what is your brother's name?'

Mime

- *Mime different ways of moving through a hole:* 'hide in a hole' – curl up small and then emerge slowly out of the hole; 'stuck in a hole' – pretend to be crawling through a hole and get stuck; 'fall in a hole' – mime falling into a hole; 'lost in a hole' – go through the hole and pretend to get lost. What is in the hole?
- *Pass the holey prop:* Sit in a circle and pass round a prop such as a hoop, ring, hair bobble or loop of string. Ask the children to mime using the 'holey' prop in different ways. Can the others guess what they are pretending it is?

Drama skills

- Use hoops to represent entrances to holes. As the children climb through they find themselves somewhere else. Ask them to make up where they are. Just choose one thing to start with that has changed such as temperature, time of day, mood etc. Then extend the improvisation and use as the start of a fantasy story.

Role-play
Animal hole or burrow

Set up: Make a dark area using drapes with a play-tunnel to crawl through as the entrance. Use some home corner furniture and props to create an animal home in the hole.

Roles: Animals such as rabbits, badgers, moles, foxes etc.

Stories: A new hole to furnish and clean; hold a 'hole warming party 'in the hole; the animals home is under threat from the farmer or building work etc.

Holidays

Drama games

- *Beachcombing:* Pass items around the circle found on the beach and ask children to pretend they are something else. A shell becomes a hat or a mobile phone, a piece of wood is shaped like a knife, the seaweed becomes a brush for sweeping the floor. How many different uses can the children invent for each item?
- *Hot seats:* Invite children to talk for half a minute about a favourite holiday. Record or film their efforts. This is a good game for developing children's confidence in speaking to a group. Ask the children to choose alternative subjects to talk about.

Mime

- *A walk in the sand:* Mime walking on different surfaces and move appropriately. For example soft sand – slippery, feet sink into the sand; hot sand – starts to get hot from the sun, take small quick steps on tip-toes to avoid being burned; pebbles – difficult to walk on, hurt feet, and fall over; rocks – clamber over the spikey shapes, and scratch or cut feet; wet sand – draw a picture or write your name with your toes.

Drama skills

- Go on a journey to the beach. Act out the sequence of events involved. Pile into the car with lots of equipment. Find a good place to sit or set up camp. Apply suntan cream, change and go for a swim in the sea. What if the water is cold? Eat picnic food and read a book. Build a sandcastle, go fishing in the rock pools, eat an ice-cream. As the sun goes down, pack up and go home.

Role-play

A travel agents

Set up: Tables and chairs, telephones, computer screens, diaries, lots of holiday brochures and posters, tickets, passports.

Roles: Travel agents, customers.

Stories: Organising a dream holiday; holiday-makers come in to complain very loudly; a rude customer on the phone keeps interrupting, and so on.

Houses and Homes

Drama games

- *Four corners:* Talk about different sorts of homes that animals live in. Choose four types of animals' homes such as a nest, burrow, hole in a tree, or cave and draw or label the four corners of the room. Ask all the children to move around the room until you call out the name of an animal, such as a sparrow, squirrel, rabbit, badger, bee, bat, bear, etc. The children must then choose which home to run to.

Mime

- *This is the house that Jack built:* Use this rhyme to mime building a house together.
- *Who's that at the door?:* Ask for a volunteer to come and knock on the door, then come in the room and mime a character for the others to guess. Use dressing-up clothes and props to develop mimes.

Drama skills

- *The Three Little Pigs:* Retell the traditional story. Divide into scenes and act out the story. Talk about the characters of the three little pigs. How could they vary from each other? Introduce 'hot seating', i.e. speaking and answering as if they were a character in the story. Rewrite the ending.
- *The haunted house:* Improvise a drama about a haunted house. Introduce characters that are brave, lazy, scared, bossy, and small. They dare each other to visit the big, empty house and hear lots of strange sounds. How do the different characters react?

Role-play

Estate agents

Set up: Tables and chairs, computers, telephones, house details, local newspapers, a diary, bunches of keys.

Roles: Manager, estate agents, customers: both buyers and sellers.

Stories: Visit a house that is not for sale by mistake; be late for an appointment; encounter a very fussy or angry customer; the house is falling down, and so on.

Journeys

Drama games

- *In my shoes:* Make a collection of different types of shoes. Put them in the middle of the circle and invite children to try them on and assume the character of someone who would wear them. Ask them to invent a line of dialogue to introduce their character. Can they go on an appropriate journey? For instance if they choose Wellington boots they can pretend to be a gardener walking round his garden, or ballet slippers a dancer who is late for a performance.

Mime

- *Follow my leader:* Play follow my leader games around the room. Make the path change between straight, bendy, and sharp turns. Vary the levels at which children move by asking them to dip under branches, crawl through tunnels and climb up hills, over stiles and other obstacles. Walk alongside streams, wade through water and balance across bridges.
- *Magic carpet:* See the topic of Flight.

Drama skills

- *Journey dilemmas:* Improvise journeys and introduce dilemmas or problems for the children to solve such as: getting lost, being late for an important date, forgetting something important, leaving someone behind, getting stuck in a traffic jam, a cancelled train or a car accident.

Role-play

A garage

Set up: Sit-and-ride vehicles, tools, overalls, spare parts and tyres, a petrol pump made from a painted cardboard box with dials and a hose, a radio, cash till, car magazines, a telephone. Set up a garage shop with newspapers, flowers and confectionary. Provide props for the car wash: buckets, sponges, shammy leathers, hoses etc. Build a breakdown truck and use a tow rope to rescue broken-down vehicles.

Roles: Manager, mechanics, petrol-pump attendant, salesperson, customers etc.

Stories: The car is not ready on time for the customer; there is trouble mending the car; the garage runs out of petrol; a customer's car is scratched; a rude customer, and so on.

Light

Drama games

- *Pass the torch:* Sit in a circle and pass a torch around the circle as you sing this song to the tune of *Three Blind Mice*:
  > Pass the torch X2
  > Switch on the light X2
  > Can you pull a silly face?
  > Can you pull a funny face?
  > Can you pull a scary face?
  > Give me a fright!

  The child holding the torch at the end of the song can shine the light up under their chin and pull a face of their choice.
- *Warm up 1, 2, 3:* Try this warm up game. Ask children to stand in a circle and then walk round. Tap the tambourine once to change direction. When they are confident with this add a further instruction: tap the tambourine twice to change speed. Finally, tap three times to change the level at which they are traveling. This final stage is quite challenging for young children so don't introduce it until they are ready.

Mime

- *Blind pairs:* Talk to the children about what it must be like not to be able to see anything, including light. Ask them to work with a partner. Blindfold one of the children and ask their partner to lead them around the room taking care they don't trip or bump into anything or anybody.

Drama skills

- *Dark and light:* Choose different activities to try with eyes closed or wearing a blindfold and then repeat 'in the light' – putting on shoes, writing their name, having a drink, or walking across the room.

Role-play

A dark cave

Set up: Dark drapes, curtains and screens to create a 'cave' space, home corner furniture, cushions, luminous stars and shapes, lanterns, torches, lamps, soft-toy bears, bear masks or dressing-up suits.

Roles: Cave dwellers, bears, explorers.

Stories: A midnight snack in the cave by lamplight; explorers looking for treasure find a cave. A group of children are playing by the sea, the tide comes in and they hide in the cave. A family of bears hibernate in the cave waiting for spring to come. Act out the story of *Can't You Sleep, Little Bear* by Martin Waddell.

Machines

Drama games

- Sit in a circle and pass machine sounds around the ring. Use machine voices to greet each other and ask questions.
- *Pass the phone:* Pass a mobile phone around the circle until it rings. Ask for a volunteer to make the ringtone. The child who pretends to answer it must improvise a conversation or argument.
- *Phone moods:* As before pass a phone around the circle. Invite children to pretend that they are hearing bad, happy, sad or worrying news. How will they respond?

Mime

- *Remote control:* Ask children to find a space in the room. Explain that you have the remote control and that they are machines and have to do what you say, immediately! Give one-word instructions such as: walk, forwards,

backwards, jump, hop, up, down, stand, sit, spin. Invite confident children to have a turn at being in control of the machines.

- *What's my machine?:* Invite children to mime using a machine for the others to guess. Try miming using a mobile phone, camera, computer, calculator, kettle, toaster, mixer, hairdryer, hoover, drill, mower, and so on.

Drama skills

- Ask each child to create a machine sound, and then add a repeated movement like a machine. Ask them to work with a partner and combine their sounds and movements together. Go round and switch the machines on and off.
- *Group machines:* Join the children together to form a group machine. Make the movements interact and synchronized. Give each child a number and ask them to start making their sound and movement when they hear their number. Use the numbers to start and stop the group machine.

Role-play

Factory floor

Set up: Make a giant machine using a climbing frame, slide, A-frame, ladders, hoops, balance beam etc. Add a conveyor belt and buttons to press.

Roles: Foreman, workers, objects going through the machine.

Stories: The workers go on strike; the machine changes speed, either very fast or slow, and then breaks down; vary the finished product that each machine makes from cars or toys to baked beans; somebody gets hurt in the machine.

Materials

Drama games

- Display a collection of different items made from a variety of materials. Ask one child to choose one item, place it into a bag without being seen by the others, and then describe it to the rest of the group by touch. Can they guess what is in the bag? See the topic of Senses.
- *Sticky feet:* Improve children's posture and deportment using this fun game. Pretend to apply super glue to the children's feet so that they are stuck firm to the floor. Stick them in different positions, close together, far apart, hip width apart: which is the most stable?

Mime

- *Sculptors:* Ask children to work with a partner, one is the sculptor, one the clay. Invite them to sculpt the clay into sculptures using different materials, wood, clay, metal, junk, sand etc. Once the sculpture is ready can the others guess what it is?

Drama skills

- *Waxwork museum:* Describe to the children a museum or collection of statues and then ask them to model themselves into a statue or use some statues from the Sculptors game above. Start with various characters from stories or different sports. Walk round and see if you can identify what the children have become. On a signal ask the statues to come alive, and act or mime their character or sport to help them be identified.
- Improvise a drama about the museum where one day or night the waxworks all come to life by magic and scare the visitors or the curator.

Role-play

A builder's yard or building site

Set up: Bricks, wood, sand, gravel, buckets, tools, spades, pipes, rulers, measures, striped tape, string, ladders, cones, wheelbarrow, signs, spirit levels, paint brushes, tins of paint, overalls, hard hats, boots, plans, stripey workman's hut, wall paper, paint charts, tea-making equipment etc. Home-made bricks made from small cardboard boxes stuffed with newspaper that are sealed and wrapped in brown paper.

Roles: Manager, foreman, architect, builders, workmen, customers.

Stories: Somebody gets injured on the site when tools are left lying around, work is delayed. Poor quality materials and a wall falls down. There is a mistake in the plans and the door is in the wrong place. Celebrate the opening of the new building. Act out some *Bob the Builder* stories

Minibeasts

Drama games

- *Life cycles:* Talk about the different stages of growth involved in a lifecycle of a frog or butterfly. Introduce terms such as egg, caterpillar, pupa, butterfly. Sit in a circle and ask children to say the words in the correct order around the ring. Add actions. Anyone who gets it wrong has to sit out.

Mime

- *Minibeast Moves:* A fast-moving miming game in the style of 'Beans'. Call out names of different minibeasts for children to mime. Start with a butterfly – flying around the room, a bee – buzzing as they fly, a worm – wiggling on the ground, a caterpillar – crawling on all fours. When the children are confident with these add more such as a ladybird – walking daintily on tip-toe, a flea or frog – jumping, a snail – sliding around the floor, a spider – join with a partner on all fours to create eight legs, and finally a centipede – all join together in a long line with hands around waists.

Drama skills

- *Little Miss Muffet:* Act out the nursery rhyme and help children to show different emotions. Start quite happy when Miss Muffet is eating her supper and then change to 'fear' when the spider arrives. Can the children think of a way of changing the rhyme in order to change the moods?
- Improvise a drama when the children are magically reduced in size and find themselves in a garden where they encounter lots of different minibeasts? What would they see, hear, feel, etc? How could they get back to their original size and world?

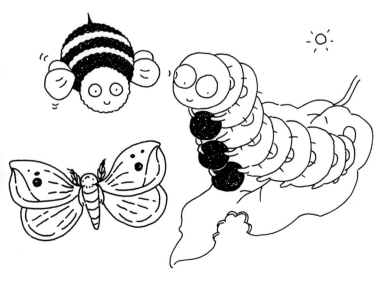

Role-play

Minibeasts' garden or jungle

Set up: Trees and plants made from corrugated cardboard, and green crepe paper, hanging creepers, butterfly nets, holes in tree trunks, green or brown mats, soft-toy birds and plastic bugs. Make an entrance way through a curtain made of green streamers or garden netting. Make home-made bugs made from cardboard circles, pipe-cleaners, net wings, buttons, etc.

Roles: Explorers, visitors, minibeasts.

Stories: Discover a brand new minibeast in the jungle; get bitten or stung by a bug; imagine life as a tiny minibeast, escaping from predators and humans!

Night

Drama games

- *Chinese night whispers:* Sit in a circle and pass a nighttime message around the circle very quietly. Does it stay the same?
- *Drama in the dark:* Make the room as dark as possible and let the children get used to it. Ask them to stand in a space. Play peaceful music to set the scene for 'night'. Invite the children to move slowly around the room, taking care not to bump into anybody. Help any nervous children by holding their hand or letting them go round with a friend. Ask them to pretend to be nocturnal creatures such as foxes, cats and badgers creeping about, or bats and owls flying around. End the game by curling up on the floor and having a quick sleep!

Mime

- *What's my line at night?:* Invite children to mime doing different jobs that happen during the night such as nurses, doctors, milkmen, postmen, lorry drivers and police officers.
- *Throw that feeling:* See the topic of Friends.

Drama skills

- *Night characters:* Ask children to work with a partner, each with a prop, and improvise a bedtime adventure. Suitable props could be a torch, umbrella, toothbrush, story book, teddy bear, toy, mug, candle, pillow, telescope etc.
- Improvise a night drama. Choose some characters to go on a night walk, hear some strange sounds, become very scared. What do the noises turn out to be?

Role-play

Home corner at night

Set up: Home corner furniture, black-out curtains or windows, pyjamas, dressing-gowns, slippers, cocoa, biscuits, a television, bed, duvets, pillows, lamps, torch, toothbrushes, storybooks, clock.

Roles: Family members, children, babysitter, visitors etc.

Stories: Different bedtime situations: somebody doesn't want to go to bed; someone refuses to brush their teeth or insists on another story; a child plays up to the babysitter. Create a peaceful bedtime scene with a child cuddled up for a cosy story when they hear a strange noise downstairs, is it a burglar? Act out the story of *Peace At Last* by Jill Murphy.

Numbers

Drama games

- *Buzz 1 2 3:* Stand in a circle and count '1', '2', '3', 'buzz' round the ring. Those who say 'buzz' have to sit down. Who is left standing at the end?
- *Huggy 2, 3, 4:* This is a great warm-up game for breaking the ice! Ask children to move around the room taking care not to touch anybody else until they hear you shout out 'Huggy 2'. This is the signal to find a friend to hug. When they hear the shout 'Huggy 3' or 'Huggy 4' they have to get into a group of three or four and have a group hug!
- *Musical numbers:* Move to the music and when it stops find a partner and make the shape of a number in the air or on the floor. Decide on which number to make before the music begins again.

Mime

- *Moving numbers:* Give each child a number from 1 to 4. Ask all the number 4s to move across the room as though they are carrying something very heavy. Then ask all the number 1s to skip across the room. Ask all the 2s and 3s to make wheelbarrows and move together across the room. Think of different ways for each set of numbers to move.

Drama skills

- *Maps:* Help children to draw maps of islands on squared paper and add interesting places to visit. Number the squares along the top and side and select co-ordinates to choose a route around the island. Ask children to create a drama as they explore their maps.

Role-play

A row of terraced houses

Set up: Multiple sets of home corner furniture to create a row of houses, with separate front doors and different house numbers. Add sets of matching coloured props for each house and letters with the house numbers on. A postman hat and bag, sit-and-ride toys, boxes to deliver to houses.

Roles: Families, neighbours, a postman, delivery driver, visitors.

Stories: Families visiting their next-door neighbour; a letter or box delivered to the wrong address; a noisy party in one house, upsetting the neighbours.

Opposites

Drama games

- *Opposites word game:* Two children challenge each other to be the quickest to say the opposite to any word you say. The winner then has to be challenged by a new child.
- *No and Yes!:* This is challenging for young children. Ask a question that requires a yes or no answer and they must say the opposite answer, for instance 'are you a boy?' or 'have you got three legs?' Try not to laugh and that makes it even harder!

Mime

- *Mirror opposites:* See the topics of Food and Ourselves. Ask children to work with a partner. Whatever the partner mimes the mirror has to do the opposite! So if one lifts up their right hand and waves, the mirror does the same, but with the left hand etc.
- *Happy and sad masks:* Make some opposite mood masks such as happy/sad, angry/calm, scared/brave. Let the children wear the masks and mime a character to fit the mask.

Drama skills

- Ask children to work with a partner and make up a drama about two opposite characters such as a noisy boy and a quiet girl, a giant elephant and a tiny mouse, and a good caterpillar and a bad slug.
- *The Hare and the Tortoise:* Act out the story using as many opposites as possible.

Role-play

Two opposite areas, e.g. hot and cold

Set up: Hot area: use a yellow mat for the sand, swimming costumes, sun hats, sun glasses, sun-cream, beach towels, buckets and spades, a blue mat for the seawater, a cold snack of ice-creams and iced drinks. Cold area: use a white mat for the snow, Wellington boots, scarves, hats, gloves, mittens, skis, soft-toy polar bears, penguins, seals, a warm snack of cocoa and hot, buttered muffins.

Roles: Children, families, life-guards, holiday makers, polar explorers, Inuits etc.

Stories: In the hot aread reenact a family holiday on the beach; learning to swim; safe sun-bathing; a lost child on the beach; a sand-carving competition. Then in the cold area get children to explore the North pole, walk in the snow, and build a snowman.

Ourselves

Drama games

- *Catch my name:* Sit in a circle and say your name. Then throw a bean bag or soft toy to a child. They say their name and then throw to another child. Continue until everyone has had a turn.
- *Name game:* Go round the circle helping the children to invent alliterative names for themselves such as Jumpy John, Angry Adam, Moaning Megan, Happy Harriet and ask them to make up an action to go with their new name.
- *Head to toe:* This is a riotous warm up game. Ask children to move carefully around the room. Shout out two body parts e.g. knees and head, and the children must join their knee to someone else's head as soon as possible!

Mime

- *Throw that feeling:* See the mime game suggested in the topic of Friends.
- *What's my line:* Talk about different jobs that the children's parents, families and friends do. Invite them to mime a job for the others to guess. See also the topics of Night and Growth.
- *Mirrors:* Work with a partner and point to different parts of the face, body etc. as the mirror tries to copy. Sing 'head, shoulders, knees and toes' for the mirror to copy. See activities suggested in the topics of Food and Opposites.

Drama skills

- Improvise a drama in a film or photographic studio. Talk about the characters: the camera operator, sound man, make-up artist, director, actors, agents, and so on. Record a popular current advertisement from the television or a traditional story. What could go wrong during the filming?

Role-play

Doctor's or dentist's surgery

Set up: Doctor's surgery – nurse's uniform, white coats, doctor's kit, stethoscope, prescription pad, phone, diary, computer, posters about keeping healthy, empty pill bottles, bandages, sling, pretend casts, wheelchair. Dentist's surgery

- uniforms, a special chair, bib, toothbrushes, charts showing teeth and how to look after them. A waiting room with magazines, comfy chairs and reception area.

Roles: Doctor, nurse, dentist, receptionist, patients.

Stories: Go into the doctor's groaning and in pain, keep telling the doctor a different part of your body hurts; the first visit to the dentist; a doctor can't work out what is wrong with the patient.

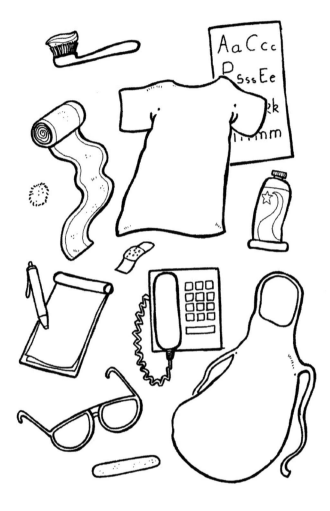

Patterns

Drama games

- *Sound circle:* Sit in a circle and choose two or three body sounds to create a repeating pattern, for instance 'clap, tap knees, clap tap knees'. Pass the pattern around the circle. Try again with vocal sounds such as 'tick, ss, wow'.
- *Echo clapping and singing:* Clap a simple rhythm pattern and invite children to echo you. When they are confident ask for volunteers to lead and clap the patterns. Try singing a simple two-note pattern based on the cuckoo call and invite the children to echo.

Mime

- *Body patterns:* Ask children to work with a partner. Stand facing each other and create patterns such as stretch up, crouch down, clap hands, turn around. Can they make up their own pattern and demonstrate it to the group?
- *Letter shapes:* Ask pairs of children to use their bodies to create the shapes of their initials. Remind them that they can stand up, sit down or lie on the floor. Try again with numbers.

Drama skills

- Create some physical drama with the whole group. Make patterns using children's bodies. Ask children to work with a partner to create an arch. Use shapes to create a zigzag across the room. Ask children to stand in rows and stretch arms and legs wide. What other patterns can the children make with their bodies?

Role-play

Art gallery

Set up: Mount and display lots of artwork showing patterns done by the children or artists. Patterned wallpaper, patterned posters and tickets, a ticket office with a phone, computer, cash till, money, wrapping paper and bags. Make a catalogue for the gallery containing the names of the paintings, artists and prices.

Roles: Visitors, sales assistant, artists, manager.

Stories: Opening night of a new show at the gallery; a very nervous artist; a valuable work of art is stolen; a competition to choose the best patterned artwork; a gallery is under threat of closure; a rude visitor doesn't appreciate the show.

Pets

Drama games

- *Introductions:* Sit in a circle and take turns to introduce yourself and then talk to the group about your pet. For example 'my name is Judith and I have two cats called Smudge and Ronnie and a rabbit called Flopsy.' See also the topic of Colours.

Mime

- What's my pet? Choose different pets to look after in a mime: take your dog for a walk or throw a ball for it to fetch; stroke the cat; play with the hamster in its wheel or ball; let your mouse run from hand to hand; feed your bird. Can anyone guess what sort of pet you have?

Drama skills

- Improvise a drama about a new pet, for instance 'the snuffletoise'. You are asked to look after this new pet for someone. Describe it and its needs. Agree to look after it, nervously, and do your best. Resolve the drama by pretending that an alien turns up to ask for his snuffletoise back!
- *The lost pet:* Ask the children to work with a partner and create a poster about a lost pet. Take turns to phone the police to report the lost pet. Ask them to sit back to back with one playing the owner and the other the police officer.

Role-play

A pet shop

Set up: Soft-toy animals, baskets, cages made from flexible, plastic straws or cut-out boxes, magnetic fish, pet food, cash till, money, posters, pet toys, hutches, collars, name tags etc. Make home-made hamsters by stuffing tights with buttons sewn-on for eyes.

Roles: Petshop owner, sales assistant, customers, pets, animal breeder.

Stories: A rare animal escapes in the shop; an awkward customer can't decide which pet to buy; a best-in-show pet competition; buying a new kitten.

Pirates

Drama games

- *What's the time Captain Hook?:* Play this game in the style of 'what's the time Mr Wolf?' Choose a child to be Captain Hook and stand facing away from the rest of the children. They all chant the question and he replies with a time such as 'three o'clock', and the others all take three steps nearer to him. The aim is to get close enough to touch Captain Hook without him catching you. If he calls 'it's time to walk the plank' he is allowed to turn around and try to catch a child to take his place.

Mime

- *Walking the plank:* Give each child and yourself a folded-up piece of paper. First time round make sure that you get the paper with a black spot marked on and have to walk the plank. Mime being scared as you walk along a narrow plank of wood and then jump into the deep water. Play again and let some of the children walk the plank.

Drama skills

- Help the children to draw treasure maps on squared paper with an X marking the spot where the treasure is hidden. Improvise a drama about searching for treasure on an island. Talk about the different characters that might be looking for the same treasure. Will the pirates co-operate with each other? What happens when they find the treasure?

Role-play

A pirate ship

Set up: Use wooden bricks, soft building bricks or a climbing frame to create a pirate ship. Add ladders for climbing on board and a balance beam for a plank to walk along. Use hoops or old tyres as port holes. Hang up a large sheet for the sail and add ropes and string. Point out any hazards and explain to the children that they will have to be very careful not to trip when playing on the pirate ship.

Roles: Pirates, sailors, mermaids, Captain Hook, Captain Jack Sparrow, prisoners.

Stories: Talk to the children about any pirate stories they are already familiar with that could be role played in the pirate ship. Sail to a treasure island, chase another ship, make someone walk the plank, run out of healthy food at sea.

Puppets

Drama games

- *My friend Albert….:* Take along a large hand puppet, preferably with a moving mouth, and introduce it to the children. It doesn't have to be called Albert! Help them to befriend the puppet and gain confidence so they can speak to it without embarrassment. Use the puppet at each drama session to introduce new ideas and challenges. Children sometimes find it easier to relate to a puppet than they do a new adult.
- Make a collection of finger or hand puppets and let each child choose one to use. Make your puppet speak to the group. Take turns for children to introduce their puppet to the group. Ask them to choose a partner and invite their puppets to have a conversation.

Mime

- *Come to life:* Ask the children to find a space in the room and stand still like a puppet made of wood, plastic or fabric. On an agreed signal ask them to slowly 'come to life'. Practise this transformation gradually with each part of the puppet slowly tingling, then wiggling, then moving more purposefully until the puppets are dancing.
- *Teaching Albert:* Choose an activity and mime instructions for the puppet Albert (see above) to follow such as making biscuits, getting ready for bed, or buying a new pair of shoes.

Drama skills

- *Home-made puppets:* Encourage children to make their own simple finger puppets from paper or felt and to improvise dramas and stories using them in a small group.
- *Pinnochio:* Talk about the story of Pinnochio. Relate this to the mime game 'come to life' above. Act out parts of the story.

Role-play

A puppet show

Set up: Lots of different puppets, a mini-theatre made from a large cardboard box with the front cut out and a curtain added. Alternatively, hang up a sheet or curtain over a washing line and let the children stand behind and hold the puppets up over the top.

Roles: Puppeteers, audience, ticket sellers, programme designers etc.

Stories: Use puppets to tell any traditional story or improvise new ones. Puppets could get lost or start to misbehave; a noisy audience heckling; the show starts late, or the curtain won't open, and so on.

Recycling

Drama games

- *Prop box:* Put together a collection of items that could be used as props such as a telephone, handbag, bunch of keys, plane tickets, wallet or purse, birthday card, newspaper, camera, torch, slipper, egg cup, sun cream etc. Put them all in a box and label it the 'prop box'. Invite children to take turns to pick an object from the box and think of a sentence or line of dialogue including the name of the object.

Mime

- *Pass the prop:* Sit in a circle and choose an item of junk to pass round. Each child has to mime using the 'junk' prop as though it was something else. Can anybody guess what they are pretending to use? Try using a cardboard tube, saucepan lid, wooden spoon, feather, aluminum foil tray etc. See the topic of Holes.

Drama skills

- *The story of a newspaper:* Talk about where paper comes from and trace the journey from a tree being chopped down, to a saw mill, to being flattened into pulp, being made into paper and printed on. Then finally being read and recycled. Help the children present the information in a drama showing each stage of the process.

- *Prop box dramas:* Extend the prop box game (see above) into improvised dramas. Ask a pair of children to each pick a prop by either design or chance and to make up a drama using them. Start with 'action' and ask them to stop when you say 'freeze'.

Role-play

Recycling centre/Second hand shop

Set up: Plastic boxes and bins labeled for newspaper, cardboard, drink cans and tins, clothes and shoes. Make a bottle bank from a huge cardboard box painted green with holes cut in and let the children drop empty plastic bottles into it. A broom to sweep up the mess. Dressing-up clothes and shoes in different sizes to sort.

Roles: Manager, customers, sorters, cleaner.

Stories: All the boxes are full, where will the rubbish go? A lazy customer puts the wrong things in each box; the recycling centre is threatened with closure, get the children to organise a campaign to save the centre via a demonstration and produce placards.

Senses

Drama games

- *Long distance news:* Invite the children to take turns to share some news with the group. Line up all the children at one end of the room and ask them to project their news across the room to you. Encourage them to speak slowly, clearly and as loud as they can without shouting.
- *Chinese whispers:* Sit in a circle and pass a secret message around the ring. Does the message get through unchanged?
- *Feelie bag:* Show children your prop bag and a collection of useful props. Put a prop in a cloth bag and ask a child to describe it to the group by touch only.

Mime

- *Blind pairs:* See the topic of Light.
- *Which sense?:* Mime different sensory activities for the group to guess such as smelling a flower, eating food, looking at a picture, listening to music, stroking a soft toy.
- *Nice and nasty smells:* Ask children to stand up tall and practice breathing in and out. See the activity in the topic of Growth. As they breathe in ask them to mime smelling nice smells such as flowers, perfume, chocolate, cookies, and fresh bread, or nasty smells such as mouldy cheese, dirty socks, manure, wet dogs etc.

Drama skills

● Improvise a drama about a group of children who have one of their senses highly developed, i.e. one can see things happening miles away, the other can hear sounds from a great distance. How could they help each other and other people?

Role-play

An opticians

Set up: Swivel chair, eye charts, lots of glasses frames, mirrors, coloured lights or torches, a waiting area, posters, brochures, magazines, a receptionist, computer, diary, phone. Make your own eye-charts using shapes, numbers and some letters.

Roles: Optician, receptionist, patients.

Stories: A patient is worried about not being able to see properly; a fussy customer can't decide which glasses suit her.

Shapes

Drama games

● *Make a shape:* Ask children to find a space in the room and dance to the music. When the music stops, call out a shape and ask the children to make that shape with their hands or body.

● *Circle warm up:* Stand in a circle and all hold hands. Swing arms in and out, step in and out, drop hands, stretch arms up high, sink down low, stretch arms out front, to your sides, behind, march on the spot, hold hands again and walk around to the right and then to the left, stand still and shake hands, arms, feet, legs etc.

Mime

● *Rollaball:* Sit in a circle and roll a ball to one child in the ring without speaking. Just use eye contact and use a nod of the head or a wink. Then they must do the same and choose somebody else in the circle and roll the ball.

● *Pass the shape prop:* Pass a 2D shape around the circle and ask the children to mime using it as something that shape, for instance, a circle could be a wheel, plate or frisbee; a square could be a window, bag, book, etc. Try with 3D shapes e.g. a sphere could be a ball, fruit etc.

Drama skills

● *Shape characters:* Create characters in the style of the Mr. Men series and make up stories about them. Mr. Circle is round, easy going, quick, happy. Miss Square is slow, solid and reliable. Mrs. Triangle is nervous, edgy and a bit prickly.

Role-play

A shapes workshop

Set up: A work bench, carpentry tools, scissors, glue, tape, wood, cardboard, boxes, tubes, different shaped templates, junk materials etc. Plans for different shaped items such as shaped greetings cards, shaped models, jewellery made of shapes, shape pictures, and so on.

Roles: Carpenters, designers, customers.

Stories: It is a busy time in the workshop with lots of orders for different items; a special model gets broken or lost; the carpenters run out of materials; a new worker gets all the shapes mixed up; there is a shape picture competition.

Shopping

Drama games

- *I went to the supermarket:* This is a good game for concentration and listening skills. Each person adds to the list of items bought at the supermarket. See the topic of Food.
- *Who will buy?:* Choose four different shops and put up labels in each corner of the room: bakers, greengrocers, newsagents, chemists. Read out a shopping list and ask children to run to the correct shop to buy each item. Where would you buy apples, bread, magazines, oranges, medicine, cakes, toothpaste, sweets?

Mime

- *Which shop?:* Ask children to mime visiting different types of shops and buying appropriate items. Can the others guess which shop they are miming?

Drama skills

- *Five currant buns:* Act out the rhyme in different ways. Add drama by changing the type of buns or cakes, the prices, and the customers who come to the shop. Go behind the scenes and act out making the bread or cakes for the baker's shop.
- *Marketing:* Ask children to think of some news to share. Take turns interviewing each other using an echo microphone. Remember to speak into the microphone each time. Pretend to interview people who are shopping and talk about new products and offers. Create an advertisement for an exciting new product.

Role-play

Open a shop - supermarket, bakers, greengrocers

Set up: Uniforms, shelves, tables, purses, money, baskets, trolleys, cash till, bags. Suitable stock for each shop e.g. empty food boxes packed with newspaper and sealed; salt and flour dough cakes and biscuits; pretend bread; plastic fruit; fruit made from newspaper, modroc and paint.

Roles: Shopkeepers, shelf stackers, checkout staff, baker, greengrocer, customers, suppliers.

Stories: The shop runs out of items; impatient customers; a child is lost in the shop; lazy workers; customers complaining about broken biscuits or mouldy potatoes. A shopper forgets their shopping list; there is a power cut in the shop, the freezer packs up!

Space

Drama games

- *Space conversations:* Sit in a circle and ask the children to make up some alien voices and sounds and take turns to have a 'space conversation'.

Mime

- *Moon walking:* Mime different space movements such as 'taking off' in which children crouch down, count down slowly and then jump into the air like a rocket; 'in orbit', children move round in circles slowly; 'space walk',

children are astronauts floating in space connected to the rocket by a wire; 'touch down', slowly sink down onto the moon surface; 'moon walking', pretend to walk in a no gravity atmosphere with heavy moon boots.

Drama skills

- *Noises off:* Use different musical instruments and sounds to make strange sounds behind a screen. Ask the children to react and imagine what the sounds could be. Ask the children to improvise their own sounds. Make sure each situation is resolved safely i.e. the sounds turn out to have a perfectly reasonable explanation.
- *Group aliens:* Try some physical drama. In small groups, ask children to make an alien using their bodies and faces. They will need to decide how many legs, heads, and eyes their alien has! Can they create an alien language and have a conversation with another alien in the group? Organise a parade of aliens along to some spooky music such as 'Saturn' from Holst's Planet's Suite.

Role-play

A rocket landing on the surface of a planet or the moon

Set up: Use the climbing frame as a rocket. Alternatively, arrange some home corner furniture in a circle. Cover big tyres in fabric to make craters, hang black drapes for a space backdrop with silver or luminous stars and planets. Convert a sit-and-ride toy into a space buggy by covering it with aluminum foil. Space suits and helmets to dress up in. Oxygen tanks made from empty pop bottles.

Roles: Captain, navigator, astronauts, aliens.

Stories: First voyage to a new planet; meet some friendly or hostile aliens; the rocket breaks down; the astronauts get lost on their way home. Act out the story of *Whatever Next?* by Jill Murphy,

Spring

Drama games

- *Waking up:* Ask children to find a space in the room and curl up small and pretend to be asleep, hibernating over winter. Gradually narrate their 'waking up' for Spring by wiggling fingers, stretching arms, yawning, sitting up, standing, stretching and finally springing awake!
- *Wake up voices:* Use this echo game to wake up voices after a long winter sleep. Ask children to echo or copy every sound you make with your voice. Swoop from high to low, roar, squeak, hiss etc.

Mime

- *Spring animals antics:* See the topic of Animals. Concentrate on Spring animals such as rabbits hopping, frogs jumping, caterpillars crawling, lambs bouncing, birds flying, etc.
- *The very hungry caterpillar:* Mime the story of the caterpillar as he eats lots of different food and then changes into a pupa and finally a beautiful butterfly.

Drama activities

- *The magic egg:* Ask the children to work with a partner or small group to create a drama about time travel. Give each group a small shaky egg and explain that they are 'magic' and once shaken enable the children to travel backwards or forwards in time to another time and place. Help them to make up adventures and show the rest of the group.

Role-play

Spring cleaning the home corner

Set up: Home corner furniture; lots of cleaning equipment e.g. cloth dusters, feather dusters, polish, brooms, bowls of soapy water, sponges and cloths; aprons; washing up liquid and bowl to wash all the cups, saucers, bowls and plates; party decorations and fresh flowers.

Roles: Children, cleaners, families, visitors.

Stories: Preparing the house for a special celebration; somebody refuses to help clean the house; somebody keeps getting in the way; some marks and dirt just won't come off!

Summer

Drama games

- *Take your voices for a walk:* Ask the children to change the pitch of their voices in response to some simple pictures (see diagrams).

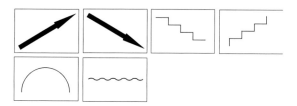

- Develop speaking voices by practising saying some summery tongue-twisters such as 'she sells sea shells on the sea shore', or 'Peter Piper picked a peck of pickled pepper'. Try these new ones: 'sing seaside songs in the season of summer' or 'nice icy ice creams and soft strawberry sundaes'. See the topic of Colours.

Mime

- *Feeling hot hot hot!:* Ask the children to mime walking through the desert, slowly stepping on the hot sand, getting slower and slower, dripping with sweat, needing a drink and collapsing onto the sand.

Drama skills

- Improvise a drama about a picnic in the park. Talk about the characters. Prepare the picnic by writing a list of food and then pretend to make the sandwiches and drinks and pack them into a picnic basket. Mime the journey to the park and set up a picnic spot. What could go wrong at the picnic? Someone could get lost, stung by a wasp, or the weather turn bad, dad remembers he has to go to work. What could go right?

Role-play

Ice cream van or stall

Set up: Convert a screen and chairs into a van, attach cardboard wheels to the side and add a striped awning. Make ice creams from cones of cardboard and tissue paper. Cash till, money, lolly wrappers, posters, price lists etc. Record a jingle for the ice cream van using a glockenspiel.

Roles: Ice cream seller, customers, other drivers.

Stories: There are no customers, too many customers; a rude ice cream man; the van runs out of everything, gets involved in car chase; a jingle gets stuck and won't stop.

Toys

Drama games

- *Introductions:* Sit in a circle and take turns for each child to introduce themselves saying 'my name is _____, and my favourite toy is _____.' See the topics of Colours, Pets, and Weather.
- *Circular news:* Introduce a special toy to the children and explain that as it goes round the circle children can only speak when they are holding the toy. Invite them to share their news, thoughts and feelings. If a child is reluctant to speak suggest that they pass the toy onto the next child.

Mime

- *Rollaball:* See Shapes.
- *Toy parade:* Invite the children to move around the room like different toys such as marching toy soldiers, floppy rag dolls, racing cars, bumbling teddy bears, stiff bleeping robots, jumping jack-in-the-boxes, dancing dolls, trundling Daleks, and spinning tops.

Drama skills

- *Film a toy advert:* Choose a favourite toy, or design a new toy such as 'a chocolate button machine' or a 'robotic frog', and film an advertisement for it. Talk about why children will want this toy. What can it do that is new? How much will it cost? Ask children to pretend to be in the studio with cameras filming the advert. What might go wrong?
- *Toy Story:* Talk about the Toy Story films and act out favourite scenes such as the arrival of Buzz LightYear, moving house, and escaping from the toy collector.

Role-play

A toy shop

Set up: Tables and shelves full of toys. Sort the toys into dolls and soft toys, cars and moving toys, books and puzzles, and so on. Cash till, money, sales desk, price labels.

Roles: Manager, sales assistant, customers, different toys.

Stories: Choosing the best-selling toy for Christmas; the toys come to life at night; being the last doll left in the shop;

nobody buying a particular sort of toy; a competition to name a toy and win it.

Water

Drama games

- *Co-operation:* Put out enough mats so there is somewhere for everyone to sit. Explain that the mats are islands and that all the children are living on the islands surrounded by water. When you say 'action!' they must leave the island and go fishing. Each time you say 'home' they must return to the nearest island. Pretend that a big storm comes and one of the islands is flooded. Now they must co-operate and fit into less space. Continue until there is only one island for all of them to squash onto.

Mime

- Sit in a circle and ask for a volunteer to mime using water in different ways for the others to guess. Try drinking water, making tea, washing hands, taking a shower, washing up pots, cooking, swimming, rowing a boat, having a bath, cleaning vegetables, and keeping fish.

Drama activities

- *Row row row the boat:* Ask children to sit with a partner and sing and act out the well-known song 'Row the boat'. Try the crocodile version and row quicker.

Make up new versions with different animals, for example:

Row row row the boat,
Gently round the lake,
If you see a tiger,
Don't forget to shake!

Role-play

Undersea world

Set up: Lots of blue mats, drapes and curtains, garden netting, green crepe paper and material weeds, plastic fish and water creatures, stones, shells and rocks to sit on. A treasure chest with treasure inside. A wreck of a boat made from cardboard boxes and tubes. Mermaid suits, deep sea diving equipment, oxygen tanks made from empty pop bottles, swimming costumes, masks, flippers.

Roles: Divers, mermaids, swimmers, fishes.

Stories: Searching for treasure; unfriendly mermaids causing trouble; a search party sent out for a lost diver; going fishing; catching a new type of fish or sea creature. Act out the story of *The Little Mermaid.*

Weather

Drama games

- *Introduction:* Sit in a circle and ask children to take turns to introduce themselves by saying 'my name is _____, and my favourite type of weather is _____'. See the activities in Colours, Pets, and Toys.
- *Rain dance:* Ask the children to find a space in the room. Kneel down and tap fingertips slowly on the floor to make 'pitter patter raindrops'. Tap feet quietly on the floor as the raindrops get more persistent. Stamp feet as the rain gets heavier. Jump up and down and splash into the puddles. Use some music that gets faster and louder to accompany the dance.

Mime

- *Weather Beans:* Invite children to mime different types of weather for example: for rain put up an umbrella; for fog move slowly as though not able to see; for snow shiver uncontrollably; for sunshine lie down and sunbathe; for wind whirl around; for ice bend into a spikey body shape; and for a storm stamp your feet and clap hands.

Drama skills

- *The Weather Forecast:* Set up the television studio with a camera crew and equipment. Make a camera from junk boxes and tubes and paint it black. Hang up a large map of the United Kingdom and make a set of weather symbols. Ask for volunteers to practise saying the weather forecast. Add drama such as a flood warning, strong winds or stormy weather coming.
- *Blown by the wind:* Ask children to go on a walk as you narrate the following instructions: go for a walk, carrying a coat over your shoulder; there is a slight breeze, that is getting stronger, put on your coat. As the wind gets stronger, fasten your coat and walk quicker. The wind is against you now so put up your collar or hood. A gale is blowing and walking is getting harder and harder. Each step is a major effort. And then you FREEZE!

Role-play
Campsite

Set up: Tent, sleeping bags, fold-up furniture, camping stove, cooking utensils, plates, cups, water containers, dressing-up clothes for different weathers.

Roles: Families, the campsite owner or farmer.

Stories: The tent is blown away by the wind or storm; too much sun; a stray dog on the campsite; burning all the food on the camping stove; the car running over the saucepans.

Winter

Drama games

- *Keeping warm:* Choose three or four different ways to keep warm: rub hands together, stamp feet, jump up and down, and blow on hands. Use different signals for the children to follow.

Mime

- *Jack Frost:* Ask children to move around room in a spikey way with elbows sticking out, fingers stretched and spikey, walking in zigzag shapes. Choose one child to be Jack Frost and go around the room touching the other children and freezing them into icy statues.
- *Winter weather:* Mime different types of winter weather for example mime snowflakes through spikey body shapes, for ice shiver, for snowballs curl up small, for a snow storm whirl around fast, for a snowman freeze in the shape of a snowman, and for a hot snowman sink to the floor in a puddle of water!

Drama activities

- Ask the children to work together in a group and mime making a snowman. First, put on lots of warm clothes and walk out in the snow. Roll the snow into huge balls and construct the snowman. Add buttons, eyes, nose, mouth, and clothes. How does the group feel when the sun comes out and the snowman melts?

- *Freeze frames:* Explain that you are going to show a significant moment from a rhyme or story such as Humpty Dumpty or The Snowman as a freeze frame. Practise 'frozen' positions such as running, climbing a tree, looking startled. Play action/freeze, as in the topics of Dinosaurs or Farms. Make the freeze frame like a still photograph. Ask the children to suggest what each character might be thinking.

Role-play

Winter café

Set up: Tables, chairs, a winter menu on a blackboard, utensils, kitchen equipment, uniforms, a fire, real and pretend food. Make some winter soup and fresh bread to serve at the café. Hot chocolate, hot dogs and hot drinks.

Roles: Waiter, waitress, chef, customers.

Stories: Bad weather, staff and customers are snowed in so nobody can leave the café; a rude waitress; a waiter spills some soup on the customers; a competition for the best winter menu.